Progressive Energy Policy

Series Editors
Caroline Kuzemko
University of Warwick
Coventry, UK

Oscar Fitch-Roy
University of Exeter
Penryn, UK

Progressive Energy Policy is a new series that seeks to be pivotal in nature and improve our understanding of the role of energy policy within processes of sustainable, secure and equitable energy transformations. The series brings together authors from a variety of academic disciplines, as well as geographic locations, to reveal in greater detail the complexities and possibilities of governing for change in energy systems. Each title in this series will communicate to academic as well as policymaking audiences key research findings designed to develop understandings of energy transformations but also about the role of policy in facilitating and supporting innovative change. Individual titles will often be theoretically informed but will always be firmly evidence-based seeking to link theory and policy to outcomes and changing practices. Progressive Energy Policy is focussed on whole energy systems not stand alone issues; inter-connections within and between systems; and on analyses that moves beyond description to evaluate and unpack energy governance systems and decisions.

More information about this series at
http://www.palgrave.com/gp/series/15052

Andrew Lawrence

South Africa's Energy Transition

Andrew Lawrence
Wits School of Governance
University of the Witwatersrand
Johannesburg, Gauteng, South Africa

Progressive Energy Policy
ISBN 978-3-030-18902-0   ISBN 978-3-030-18903-7   (eBook)
https://doi.org/10.1007/978-3-030-18903-7

This Palgrave Pivot imprint is published by the registered company Springer Nature Switzerland AG
The registered company address is: Gewerbestrasse 11, 6330 Cham, Switzerland

Preface and Acknowledgements

Within the space of a few years, the topic of energy transitions has itself transitioned from being the preserve of relatively few specialists to one that commands mass attention and global concern. Increasingly, we appreciate the many ways in which local choices about energy production, distribution, and consumption can have global consequences, for generations to come. Popular struggles against fossil fuel extraction and for more sustainable alternatives, however remote geographically, are directly relevant to communities everywhere. For this reason, case studies of energy transitions may be of interest to readers far beyond the country or region in question. I decided to write on South Africa's energy transition not only for these reasons, but also because no book existed on the topic that discussed its technical and political as well as ecological dimensions. While it is impossible to do full justice to each, my hope that a succinct synthesis of these facets will appeal to readerships interested in any one of them, and to specialists and non-specialists alike.

This book draws on more than two decades of writing on and engaging with South Africans about the country's history, politics, and struggles for a better life for all. More immediately, it grew out of a 2016 lecture at the University of Kwa-Zulu Natal's Centre for Civil Society on the country's proposed nuclear power programme. I want to thank Patrick Bond, the Centre's director at the time (and now colleague at the Wits School of Governance), for his hospitality in hosting me, facilitating my presentation at the Wits School of Governance in 2018, and for debating some of the issues covered here in the process. My thanks as well to the UKZN Centre for Civil Society's current Director, Shauna Mottiar; to Harald Winkler, Jiska de Groot, and Hilton Trollip of the Energy Research Centre, University of Cape Town; and to Lucien van der Walt at the Department of Sociology, Rhodes University, for hosting talks and providing useful feedback; and to the Habib, Nagdee, and Sienaert households for their hospitality during research visits.

I received useful feedback at several conferences and lectures, including the Low Carbon Energy Democracy Conference at Durham, UK, in September 2017; the IAIA Environmental Justice in Societies in Transition conference, Durban, SA, in May, 2018; the Energizing Futures conference in Tampere, Finland, in June 2018; and the World Congress of Political Science in Brisbane, Australia, in July 2018. I'm most grateful to Kate Lebow and Jeff Rudin for their helpful feedback on earlier drafts; the usual disclaimers apply.

Several other thoughtful interlocutors generously discussed various themes related to South Africa's energy politics, political economy and political ecology, including Kolya Abramsky, Des D'Sa, Keith Dumas, Happy Khambule, Heila Lotz-Sisitka, Jonathan Neale, and France Sesedinyane. I acknowledge in particular my debt to Muna Lakhani, and join many in mourning his passing. Muna's dedication to the cause of social and ecological justice (he would pardon the tautology) is as renowned as his selfless impulse to help others. He generously provided early encouragement and shared ideas; this book is dedicated to him. Finally, thanks to Thangarasan Boopalan and Sruthi Sethu of Palgrave Macmillan, as well as series co-editor Caroline Kuzemko, for their responsiveness and guidance throughout this process.

Johannesburg, South Africa   Andrew Lawrence

CONTENTS

LIST OF TABLES

CHAPTER 1

Introduction

Abstract South Africa is typical among major fossil fuel producers for having failed to substantially transition toward renewable energy (RE) sources.
Like many other countries, it shows a marked discrepancy between official government policies and commitments on decarbonisation, and actual
progress in reducing reliance on coal. Several factors suggest nonetheless
that South Africa (like many fossil fuel-dependent Global South countries)
would benefit from a more rapid transition to RE sources: these include
(1) the unreliability of coal, its fossil fuel-based power supply, constituting a major brake on economic development, particularly for rural areas
where the infrastructure is lacking; (2) this supply's associated pollution,
which has a serious impact on local and regional health and well-being;
(3) South Africa's superior clean-energy endowments; and, not least; and
(4) "resource curse" effects of coal (and the broader mining economy).
For many years, price has not been a plausible explanation for the relative
tardiness of South Africa's energy transition, nor the exclusive determining
factor. Rather, explanations need to refer to the country's broader political
(including international and global) contexts.

Keywords Energy transition · Renewable energy · Coal · Wind · Solar ·
Resource curse

© The Author(s) 2020
A. Lawrence, *South Africa's Energy Transition*, Progressive Energy
Policy, https://doi.org/10.1007/978-3-030-18903-7_1

This book addresses a deceptively simple question: why has South Africa failed to substantially transition from polluting and more expensive fossil fuel reliance to cleaner, cheaper, and more abundant renewable energy (RE) sources such as wind and solar power? Admittedly, the same question can be posed to most other countries—in particular, those with major fossil fuel industries: this "non-event" is a global, rather than local, phenomenon. Hydrocarbons continue to provide 80% of global energy supplies. Electricity generation worldwide from oil, coal, and natural gas—although declining slightly in recent years to around two-thirds of total generation—remains far higher than its level of under 52% in 1986; whereas generation from renewable sources (hydro, wind, solar and geothermal) has increased only slightly, from 19% since 1986 to about 23%. There remains substantial national variation: more than 40 countries (including at least a dozen in sub-Saharan Africa) generate most of their electricity from RE sources, including perhaps a dozen generating at or close to 100% from them; more than 30 others generate between a quarter and a half from RE sources.[1]

This question nonetheless is particularly relevant to South Africa: its total domestic electricity generation capacity is 51.3 gigawatts (GW), of which the overwhelming majority, 91.2% (46.8 GW), comes from thermal (almost all coal-fired) power stations, with the 8.8% balance (4.5 GW, of which 3.8 GW is currently operational) installed RE capacity.[2] Despite recent increases in RE capacity, consumption from RE sources has increased more slowly. As Table 1.1 shows, South Africa's RE consumption in 2015 was only a tenth of the world average; at less than 5% of generated electricity, it lags behind both Organization for Economic Cooperation and Development (OECD) and African averages.[3]

"Coal" is indeed the deceptively simple answer to this question. For several years, South Africa has ranked seventh in the world for total amount of coal mined (and third for tons per capita), first among larger economies for the percentage of electricity generation dependent on coal, and conversely, last for the percentage of electricity generated from renewable energy sources including solar, wind, and hydroelectric generation (Table 1.1). Although coal's rank among South Africa's top ten exported commodities has slipped over the past decade, it remains high: in 2016,

[1] World Bank (2018c).

[2] IRP (2018: 58–60).

[3] Nakumuryango and Inglesi-Lotz (2016).

Table 1.1 Top ten coal producing/generating countries and their RE generation, 2015

Rank	Country	Coal production, 2015 MT	Coal-generated electricity[a] (%)	Electricity from RE sources (%) (non-hydro + hydro)
1	China	3874.0	70.3	4.9 + 19.1 = 24
2	United States	906.9	34.2	7.4 + 5.8 = 13.2
3	Australia	644.0	62.9	8.3 + 5.3 = 13.6
4	India	537.6	75.3	5.4 + 10 = 15.4
5	Indonesia	458.0	55.8	4.8 + 5.9 = 10.7
6	Russia	357.6	14.8	0.1 + 15.8 = 15.9
7	**South Africa**	**260.5**	**92.7**	**1.9 + 0.3 = 2.2**
8	Germany	185.8	44.3	26.3 + 3 = 29.3
9	Poland	137.1	80.9	12.7 + 1.1 = 13.8
10	Kazakhstan	108.7	71.6	0.2 + 8.7 = 8.9
	WORLD	**8277.9**	**39.2**	**6.8 + 15.9 = 22.7**

[a]World Bank (2018a, b, c)

South Africa's sales of 69 million tons of coal, at $4.57 billion, were its fourth most valuable mineral export, after gold ($20.7 billion), diamonds ($10.9 billion), and platinum ($8.53 billion), contributing 4.5% of export revenue.[4] Notably, however, South Africa's local sales of roughly 183 million tons of coal generated levels of revenue more than 20% higher than did coal exports.[5]

Price, however, is not a plausible explanation for the relative tardiness of South Africa's energy transition. For several years, RE sources' cost has fallen ever lower in comparison to coal generation—three years after the introduction of large-scale RE generation, they were 23% cheaper than new coal-based generation.[6] Since then, they have become 40% cheaper.[7] Although at R0.62/kWh, South Africa's cost of wind and solar generation

[4]OEC (2018).

[5]Solomons (2017).

[6]Walwyn and Brent (2015).

[7]From 2011 to 2015, South Africa's LCOE (i.e., levelised cost of energy—averaging costs over the source's life cycle) per kWh of solar-derived electricity fell from R3.65 to R0.62; and for wind, from R1.51 to R0.62. The cost of "new coal" from the two independent coal producers has been bid at R1.03. The projected costs of the new giant coal-powered stations, Medupi and Kusile, are R1.70 and R1.91 respectively. Mudavanhu and Rudin (2018).

per kilowatt hour is still higher than RE costs in other middle- to low-income countries,[8] there is no sign that the trend of RE electricity costs falling increasingly below those from fossil-fuel generation will let up in the foreseeable future. The latest available authoritative figures from Lazard show that the costs of unsubsidized, large-scale business utility-level solar and wind power fell in 2017, continuing the trend over the past decade, to levels that are now more than 50% below those of 2013 and thus lower than coal and gas generation costs as well as capital costs.[9] The latent potential to increase the diversity of supply options would also strengthen energy security for households, firms, and for the country as a whole. Fuller explanations for South Africa's energy politics rather should refer to—and can deepen understanding of—the country's broader political context. They also may help to illuminate barriers to energy transitions elsewhere, and in so doing, contribute to broader theoretical and policy debates about how such transitions can and should occur in the contemporary global context.

Nor does the country lack official commitment to reducing emissions levels. Its comparative lack of progress stands in marked contrast to the country's official statements and commitments concerning emissions reductions. It was a signatory to the Kyoto Protocols in 2002, although as a non-Annex 1 country, this entailed no emissions reduction obligations from 2008 to 2012. In 2009, ahead of the 2011 COP17 in Durban, then-President Zuma outlined an ambitious trajectory for achieving these reductions, albeit subject to international financial assistance. His cabinet approved the National Climate Change Response White Paper, which included a proposal for a carbon tax that was to have come into effect as early as the following financial year. The government's National Development Plan envisions the "transition to an environmentally sustainable, climate-change resilient, low-carbon economy" to be well underway by 2030. Any current assessment of progress to this end is necessarily provisional, as the first of the two intervening decades has not yet ended;

---

[8]IRENA (2018: 49).

[9]Lazard's (2018) estimates for South Africa include price advantages of wind power over combined cycle gas generation of 20–30%; of solar PV over peaking gas generation of 40–75%. It now estimates that even average *capital* costs for all forms of solar PV and wind (previously higher than average fossil fuel capital costs; PV and wind already enjoyed lower operation and maintenance costs), as well as fuel cell storage, are lower than those for coal, let alone nuclear power. These estimates do not incorporate environmental regulation costs or potential social and environmental externalities; doing so would make RE generation even more competitive. See Lazard (2018).

but progress thus far has been slow at best. For example, the country's proposed carbon tax—part of its commitment to the Paris Agreement on climate change—has been in the drafting stage since 2006, and will not go into effect until June 2019. While the phrase "well underway" might suggest RE generating capacity of one third or at least one quarter of the total, in fact, the Department of Energy (DoE) set a RE target in 2005 of 4% for 2013, increased by the 2010 IRP to 5% in 2020, and 9% in 2030.[10]

Given fears of free riding by large emitters in particular, it is perhaps unsurprising that South African policymakers have expressed a preference for a global accord to address climate change. "There is no doubt that a global response is the only effective and sustainable answer to this global challenge," Zuma stated in advance of hosting the COP17 meeting in Durban in 2011.[11] In its Long-Term Mitigation Scenarios (LTMS), South Africa has pledged ambitious, if voluntary, GHG emissions reduction targets of 34% by 2020 and 42% by 2025 in relation to its prior business-as-usual (BAU) projection.

Yet South Africa also has obvious, immediate interests in supporting global accords such as the COP17's best-case Scenario 1 of "Ambitious Global Agreement". Its pledges are consonant with the government's New Growth Path strategy document—released prior to COP17 by the newly formed Department of Development—which sought to prioritise "programmes and policies needed for inclusive, green growth [which will thereby] break our historical dependency on extraction and exportation of raw minerals as well as … respond to the challenge of climate change." It claimed that in so doing, it would foster the creation of 5 million jobs by 2020.[12]

The dangers of inaction grow even more apparent. The South African Risk and Vulnerability Atlas (SARVA), sponsored by the Department of Science and Technology (DST) and managed by the country's Centre for Scientific and Industrial Research (CSIR), maps climate change vulnerability in order to assist all levels of government in building climate resilience.[13] The information it provides about the likely effects of warming for the country and the region is sobering. Increasingly extreme cycles of flooding

[10]IRP (2010).
[11]ANC Presidency (2011).
[12]Ibid.
[13]SARVA (2018).

and drought threaten sustainable water resources directly, with increased runoff and rates of soil erosion. The projected growing water shortages in turn will result in greater competition between agricultural and urban water use, lower and less predictable crop yields, greater stress on livestock and woodlands, and increased likelihood of extinction for many indigenous species. An increasing frequency of extreme rainfall combined with rising temperature make the geographical expansion of the borders of vector borne diseases more likely.[14] In addition to malaria, these include cholera, as was witnessed during the floods of 2000–2001, when over 100,000 cases were reported.[15]

The dire consequences of climate-related water shortages, no less than the political and distributional struggles these consequences entailed, were dramatically illustrated by Cape Town's 2015–2018 drought—possibly the worst in its modern recorded history—and by its water rationing response. The drought, which extended beyond the Western Cape, entailed tens of thousands of job losses in the region and similar increases in numbers of people living in poverty, as well as billions of rands of lost income.[16] The municipality threatened a "Day Zero" in April 2018 when all taps to homes and businesses would be shut off because reservoirs had gotten so perilously low and in January 2018, requested residents to consume just 50 litres per day. Yet this limit was comparable to those under which more than 140,000 poor households installed with Water Management Devices have suffered since installations began in 2007; they in turn are better serviced than the nearly 15% living in informal settlements with no reliable household access to piped water.[17] The crisis also underscored the fact that the country's energy system alone was responsible for at least 5% of total water consumption, with coal generated electricity being the most water-intensive source.[18]

If the importance of achieving large-scale transitions toward energy systems with significantly lower greenhouse gas (GHG) emissions levels enjoys near-unanimity within South Africa and internationally, however, no equivalent consensus exists concerning the best or fastest means to this end. The UN's Intergovernmental Panel on Climate Change (IPCC) 2018 report recently underscored the importance of a sharp reduction of coal

---

[14]Midgley et al. (2007).

[15]WHO (2017).

[16]Smith (2018).

[17]EMG, n.d.

[18]Discussed further in Chapter 2.

generation in particular, recommending that global coal power use must fall by two-thirds by 2030 and be almost fully phased out by 2050.[19] One scenario modelling this goal, with an estimated 33–50% chance of keeping within a 1.5°C threshold, calls for a moratorium on all new coal power plant proposals worldwide. It proposes retiring all plants by 2030, and in the meantime, reducing existing plants' capacity by 3.5% per year, about the same as the U.S. achieved over the past decade.[20]

For South Africa, this would entail a steep reduction of approximately 20GW (or nearly half of a total installed coal capacity of about 42GW) from 2020 to 2025.[21] Yet this is a far more ambitious pace of transition than the South African government has ever articulated in its energy-related policies, currently described as a "peak, plateau and decline" scenario whereby coal generation, and thus emissions, *increase* over the next decade before declining thereafter. The government's recent 2018 draft Integrated Resource Plan (IRP) governing energy policy states that for the "medium-to-high" period of certainty of 2021–2030, among projects already approved, or under construction, only 12GW of older coal generating capacity would be decommissioned. Subsequently, the IRP projects an additional 16GW decommissioned by 2040, followed by another 7GW over the decade thereafter.[22]

South Africa's Intended Nationally Determined Contribution (NDC) GHG emissions target (including land use, land use change and forestry, or LULUCF) lies between 398 and 614 Mt CO_2e over the period 2025–2030. Given that South Africa's net GHG emissions for 2012 amounted to 518.3 Mt CO_2e (including LULUCF)—including a 25% increase over the 2000 level for the energy sector alone[23]—South Africa is likely to reach the upper limit of its 2030 emission reduction targets between 2020 and 2025, and approach if not exceed its mitigation target in 2030.[24] Two state-owned enterprises—Eskom and Sasol—between them are responsible for about 90% of domestic coal consumption and over

[19] IPCC (2018).

[20] Nace (2018).

[21] Ibid.

[22] These projects include 9.5GW wind, 6.8GW solar, 8.1GW gas, 6.7GW coal (including more than 5GW of already-contracted capacity from 2019–2022, and another 1GW of new coal capacity in 2023–2024) and 2.5GW hydropower. IRP (2018).

[23] DEA (2018: 11).

[24] Climate Action Tracker (2018).

half of South Africa's total GHG emissions.[25] The 2018 Draft IRP's various scenarios ambitiously forecast this level reducing to between 116 and 168 Mt CO_2e by 2040.

This may partly reflect more realistic demand growth forecasts.[26] Yet despite registering the reduced costs of PV and wind power, lower general estimates of plant availability, and 30% lower projected electricity demand, the 2018 IRP forecasts that 34 GW of coal generation would still account for almost 65% of electricity generated in 2030.[27] The anticipated emissions still represent an increase on 1990 levels (excluding LULUCF) in the range of 19–73% in 2020 and 19–82% in 2025[28]—and are thus deemed by international emissions watchdog Climate Tracker as "highly insufficient": i.e. if replicated worldwide, leading to warming by more than 3°C. Because the rate of warming is twice as great for the southern African interior than the global average—the current 1°C increase over pre-industrial levels already means 2°C and has led to a doubling of the number of "high fire danger" days per year to between 30 and 40, for example—the effects of 3°C would be utterly devastating, including a total collapse of the maize crop and the livestock industry to southern Africa.[29]

Given the sheer volume of GHG emissions by the largest emitters—more than half by China, the US, and the EU alone—for most of the rest of the world, questions of emissions responsibility are arguably tempered more by those of what is locally or regionally practical and desirable. The way that South African policymakers and other political actors make and act on these latter determinations is thus at least as salient as the country's emissions policy commitments. As is further detailed in subsequent chapters, there are equally strong local ecological and economic grounds for accelerating the transition from coal at a pace consonant with the IPCC 2018 report recommendations.

[25] Eberhard (2011) and DEA (2014).

[26] The 2010 IRP, for example, forecast electricity demand growth of 2.8% per year from 2010–2030, amounting to a near-doubling of aggregate demand from 250 TWh to 454 TWh over this period. It also increased its target for RE electricity production to 17.8GW, an increase of almost 50% from the 11.4GW put forward in the IRP prior to the public consultation process, the first in the country's history (prior to the 2010 IRP, planning was done internally by Eskom).

[27] IRP (2018: 5).

[28] IRP (2018, Figure 11).

[29] Bloom (2018).

Evidence of a shift in these policymakers' calculations can be inferred from the 2018 draft IRP's decarbonisation agenda, which is ostensibly more ambitious than any of its preceding policy documents. Neither the price of coal nor the actual or projected GDP growth rates changed so substantially over the past several years to account for this shift in forecasting. Rather, the shift reflects both the initial achievements of RE policy, the latitude that the South African government can take in energy planning, and the extent to which it responds to civil society pressures—and thus the extent to which South African energy policy is a political (and politicized), rather than technical, question. Clearly, the political calculations about the feasibility and desirability of rates of new plant construction and decommission—for both coal and other energy sources—changed from the 2010 to 2018 IRP. Explanations of the factors behind this shift—and evaluations of how much further and more rapidly the shift to RE sources can progress—are, however, fundamentally contested.

To be sure, plausible and well-known arguments favouring an accelerated adoption of RE sources apply not only to South Africa, but also to virtually all Global South countries burdened by a common set of socioeconomic ills. One recent analysis, for example, emphasizes the technical as well as economic rationality of such transitions for middle- and low-income countries such as South Africa, based on four mutually reinforcing factors. These are (1) the unreliability of these countries' fossil fuel-based power supply, constituting a major brake on their economic development, particularly in rural areas where the infrastructure is lacking; (2) this supply's associated pollution, which has a serious impact on local and regional health and well-being; (3) these countries' typically superior clean-energy endowments; and, not least, (4) "resource curse" effects, which degrade political responsiveness and transparency no less than economic growth and stability. The study thus concludes that "a clean energy transition is not necessarily an impediment to the growth aspirations of the developing world," costing only the equivalent of one to two years of global growth to achieve full transitions for all non-OECD countries.[30] Similarly, International Monetary Fund (IMF) economists have shown that around 6.5% of world GDP—including trillions of dollars each year in middle- and low-income countries—is lost in direct or indirect fuel subsidies.[31]

---

[30] Arent et al. (2017: 10–11).

[31] Coady et al. (2017).

Each of these four factors is manifestly relevant to the South African case. In terms of **(1) reliability of fossil fuel supply and access**, petroleum imports (both crude and refined) represent the largest category of imports, covering around 70% of South Africa's liquid fuel requirements and costing over \$9 billion in 2017.[32] Moreover, notwithstanding domestic sufficiency in coal, the national grid has been plagued with numerous outages and shortages. In addition to outages over a week in December 2018 and in early 2019, earlier, in June 2018, scheduled blackouts over several hours resulted from what Eskom termed "sabotage"—i.e. workers' protests over wages.[33] In 2015, the rolling blackouts were even longer in duration, causing a decline in GDP. In November 2014, Eskom implemented rolling blackouts nationwide over several days, because of a collapsed coal silo at Eskom's Majuba power station, during a period when it was already experiencing severe supply strains on the national grid; GDP contracted by 1.3%, with total losses estimated at R170 billion.[34] In 2011 and again in 2014, poor turbine maintenance at the Duvha coal-fired station caused explosions leading to significant outages.[35] All of these episodes, however, paled in comparison to the blackout of 2008, discussed in greater detail in Chapter 3, which hastened the country's slide into a protracted economic downturn. And while these episodes have periodically affected the entire country, some areas and populations have rarely, or never, enjoyed adequate and reliable access to electricity.

Regarding **(2) negative health externalities**, those stemming from South Africa's reliance on coal—and from the mining sector more generally—are also abundantly obvious. For example, the South African underground colliery fatality rate was two to eight times worse than those of the major coal economies of United States, New South Wales, West Germany, India and the UK.[36] In recent years, official figures released by the main industry lobby group, the Chamber of Mines, show an upward

---

[32] South Africa's top imports are Crude Petroleum (\$6.54B), Unspecified (\$6B), Cars (\$3.34B), Refined Petroleum (\$2.55B) and Broadcasting Equipment (\$1.87B). See OEC (2018).

[33] Cotterill (2018).

[34] Fin24 (2014) and Hofstatter (2018: 94).

[35] Styan (2015: 109–110).

[36] Leger (1991).

trend of over a hundred mining fatalities per year, after a general trend of declines from over 550 in 1995.[37]

Following diabetes, the third and fourth major causes of death are heart and cerebrovascular diseases; the sixth hypertension; and the tenth, lower respiratory diseases: for each of these, air quality is a significant contributing factor. Each year, tens of thousands of lives are lost to coal pollution alone, and tens of thousands more to indoor smoke inhalation in households lacking adequate access to electricity for cooking and heating and thus reliant on polluting fuels.[38] Eskom had agreed to the Department of Environmental Affairs' (DEA) Minimum Emissions Standards in 2015, then obtained postponements for 11 of 14 power stations on the condition that nine stations install lower-emitting retrofits—but funds for this spend, costing R134 billion, are unavailable.[39] In 2018, satellite data revealed that the Mpumalanga coal belt suffers from the world's worst concentration of NO_2 air pollution—a toxic pollutant implicated in all of the above diseases as well as in the formation of PM2.5 and ozone, two of the most dangerous forms of air pollution.[40] More dispersed, but no less dire, has been the consequence of water pollution from the coal-mining life-cycle, including acid mine drainage leeching uranium and other toxins into most provinces' major water systems (discussed further in Chapter 2).[41]

In terms of **(3) clean energy endowments**, a recent study by the US Department of Energy's Lawrence Berkeley National Laboratory finds that, with reference to a multiplicity of factors including quality of the resource (i.e. solar intensity and wind speed and frequency), distance from transmission lines and roads, co-location potential, availability of water resources, and potential human impact, the African continent's RE generating potential is several times greater than projected demand over the next several decades. It further finds that, while wind and solar photovoltaics (PV) are now South Africa's cheapest and third-cheapest form of generation, respectively, South (and southern) Africa has barely begun to

---

[37] Stoddard (2017).

[38] Holland (2017) estimates impacts in terms of early death, chronic bronchitis, hospital admissions for respiratory and cardiovascular disease, and other factors including lost productivity, at over 10,000 lives and $2.4 billion each year.

[39] Styan (2015: 113).

[40] Meth (2018).

[41] As discussed further in Chapter 2, this is one part of a much bigger problem of water systems management. See, e.g., Bond and Dugard (2008).

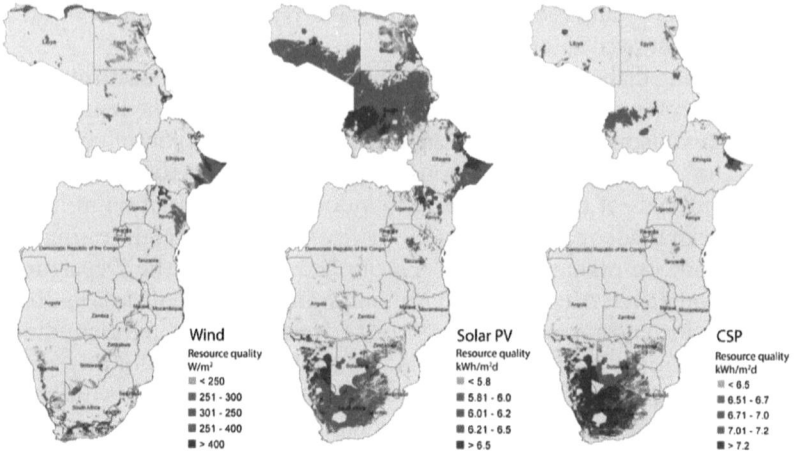

Fig. 1.1 Eastern and Southern African solar and wind generating potential (*Source* Wu et al. [2017: 3])

exploit these sources to the extent possible. Among the 21 countries in the Southern African Power Pool (SAPP) and the Eastern Africa Power Pool (EAPP)—both regions among the best-endowed in RE resources world-wide—South Africa has the greatest wind generating potential (at over 2500 TWh), the third greatest concentrated solar power (CSP) potential at over 4000 TWh, and the fourth greatest solar PV potential, at over 3000 TWh. The wind potential alone, conservatively assuming 25% capacity, is several times the country's anticipated electricity consumption of 40,000 GW for 2025. Furthermore, the potential gains from regional connectivity are even greater (see Wu et al. 2017; Fig. 1.1).

Finally, the question of **(4) resource curse effects**—a phenomenon Richard Auty famously observed over two decades ago,[42] more typically discussed with reference to oil producing countries—is one that applies not only to the prior history of several countries,[43] but to present-day South Africa as well. The term is often associated with authoritarian governments, higher levels of income and gender inequality, violence, and uneven

[42] Auty (1993).

[43] Sachs and Warner (1995: 2) and Davis (1995).

regional economic development. Minerals, specifically oil, are associated in this literature with instability, secrecy, and unaccountable governments, since petroleum rents in this view deter effective systems of taxation and their attendant modes of accountability and transparency.[44]

South Africa's dominant mining corporations certainly benefited from the country's increasingly authoritarian apartheid phase, which exacerbated higher levels of income and gender inequality, growing instability, secrecy, and lack of accountability, domestic and international violence, and uneven regional economic development. Yet despite formal democratization since 1994 mitigating chronically high levels of homicide and ending militarized engagements with neighbouring states, destabilizing episodes of resource-curse related violence persist, including outbreaks of xenophobic violence, gang violence, assassinations, and political repression such as the Marikana massacre.[45] Auty's analysis identifies five challenges to post-resource boom adjustment relevant to South Africa's macroeconomic challenges as well; as Fine and Rustomjee observed over two decades ago, for reasons still pertinent today, South African macroeconomic management "failed miserably"

> to ensure sufficient saving to cushion adjustment in the downswing (the 1980s[46] for South Africa); to ensure that tax rates and key prices do not lag behind inflation; to avoid over-ambitious large-scale investment projects; to promote non-mining tradeables; and to make adjustments to the downswing that are neither too little nor too late.[47]

These macroeconomic problems persist, signalling substantial continuity from the apartheid era. In short, there are multiple interlocking reasons—in terms of unreliability of coal supply, its attendant pollution, alternative RE endowments, and resource curse effects—for supposing that South Africa's economy and society would benefit from a swift and decisive RE energy transition.

---

[44] Humphreys et al. (2007).

[45] The 2012 Marikana massacre occurred when police responded to a wildcat strike by Platinum miners with overwhelming force, killing three dozen and injuring at least 78 strikers, making it the worst act of state-sponsored violence since the 1976 Soweto Uprising. Marinovich (2016).

[46] This stipulation applies even more to the post-global economic crisis context of the past decade.

[47] Fine and Rustomjee (1996: 248).

Yet although such rational "national interest" arguments correctly identify the benefits of energy transitions, all too often they ignore historical legacies and global economic constraints. Without showing how these contribute directly to a sub-optimal status quo, such analyses risk discounting or overlooking institutional, structural, or ideological effects.

They may also risk reifying—rather than questioning—the sharp analytical divide (also typical of most Realist international relations scholarship) between national and global interests, actors, and processes.[48] One way they do so is by framing transition processes and decarbonisation responsibilities in reductively national terms. In the case of South Africa, this framing tends to overlook the fact that the energy sector (and its closely allied extractive industries) are substantially integrated across southern Africa: Eskom historically has both imported power from and exported it to Mozambique and Zambia; it is a major supplier for Botswana, Lesotho, Namibia, Swaziland, and Zimbabwe; and it has imported power from the Democratic Republic of Congo (DRC).[49] This region also shows substantial untapped potential for further mutual gains in integrating wind and solar power—a fact that blinkered national frames are ill-equipped to address.[50]

Apart from overlooking the positive benefits of regional cooperation and further integration of energy systems, a national-international analytical divide can serve to misallocate emissions responsibility and thereby further perpetuate global climate injustices. The very means by which emissions are measured can serve to obscure transnational corporations' global value chains that outsource the most carbon-intensive industrial processes to middle- and low-income countries, thereby allowing OECD countries to absolve themselves of emissions responsibility. Taking the technology-adjusted balance of emissions embodied in trade into account, for example, reveals that for some OECD countries, imports have become more carbon intensive over time and exports less so, compared to the world economy at large.[51]

---

[48] E.g., Mearsheimer (1994).

[49] The oil and gas parastatal, Sasol, has operations even further afield, including oil and gas exploration and production in Mozambique, Gabon, Australia, and Canada; gas to liquids ventures in Qatar, Nigeria and Uzbekistan; and a US$11 billion ethane cracker and derivatives plant in Lake Charles, Louisiana (Sasol 2017).

[50] Wu et al. (2017).

[51] Jiborn et al. (2018).

Similarly, the national-international analytical divide is hardened when the effects of GHG emissions at the aggregate global level are separated from related negative externalities. Programs supported by the Kyoto Protocol's Clean Development Mechanism (CDM)—which has by no means been equally accessible among Global South states—have inadvertently intensified struggles around land in recipient countries.[52] Whereas firms in the financial and extractive industries exercise global reach in pursuit of profit, climate adaptation costs remain borne most by the states least responsible for global warming. But given the extent to which these sectors are dominated globally by oligopolistic firms—and substantially interpenetrated with each other—local challenges to their interests and domination acting alone are unlikely to succeed, unless the global underpinnings of this domination are not substantially weakened first. This is of course also true for the global supply chains of transnational renewable energy corporations as well.[53]

While it is beyond the scope of this book to provide a comprehensive review of all related research on South Africa's energy policy, it strives to highlight unanswered questions and explore the implications of contrasting, and often conflicting, approaches to energy transitions. These bear on the more fundamental question of how much capacity the country's state and political system has for embarking on a qualitatively different energy policy—and the political requisites and consequences of such a move. In elucidating this shifting landscape of South Africa's energy politics, this book seeks to contextualize and clarify policy-making approaches and their broader implications. It does so by sketching ideal-typical scenarios that are contingent: they explicitly distinguish between what is more and what is less certain; what is more and less amenable to change over the short- to medium-term; and between what is normatively desirable, and what entails unavoidable trade-offs or uncertain costs. Making these distinctions helps to make explicit provisional assumptions and causal claims.

The rest of the book proceeds as follows. Chapter 2 considers contending analytical frameworks for understanding energy transitions. Chapter 3 provides a brief overview of approaches to the SA state, highlighting in

---

[52] Newell and Bumpus (2012).

[53] As Newell and Mulvany argue, "Issues of justice will be intrinsic to whichever energy trajectory is pursued within and beyond the fossil fuel economy and they need to be better understood and anticipated by future efforts to secure a 'just transition' around questions of extraction, labour and the distribution of benefits." Newell and Mulvaney (2013: 6).

particular the role of its electricity parastatal, Eskom, and allied institutions in shaping energy policy. Chapter 4 examines the politics of non-RE alternatives: nuclear energy, geothermal energy, fracking and offshore gas. Chapter 5 considers the trajectory and fate of the country's major RE policy, the Renewable Energy Independent Power Producer Procurement Programme (REIPPPP); and Chapter 6 provides concluding reflections on alternative, neglected energy and development trajectories, suggesting how the political economy of South Africa's energy system can and should be transformed.

References

ANC Presidency. (2011). Accessible at: http://www.thepresidency.gov.za/speeches/address-president-jacob-zuma-occasion-new-age-business-breakfast-cape-town-leg%2C-cape-town.

Arent, D., Arndt, C., Miller, M., & Zinaman, O. (Eds.). (2017). *The Political Economy of Clean Energy Transitions*. Oxford: Oxford University Press.

Auty, R. (1993). *Sustaining development in mineral economies: The resource curse thesis*. New York: Oxford University Press.

Bloom, K. (2018, November 07). Interview: South Africa, climate change 'hot spot'. *Daily Maverick*. Accessible at: https://www.dailymaverick.co.za/article/2018-11-07-interview-south-africa-climate-change-hot-spot/.

Bond, P., & Dugard, J. (2008). The case of Johannesburg water: What really happened at the pre-paid 'Parish pump'. *Law, Democracy & Development, 12*(1), 1–28.

Climate Action Tracker. (2018). *South Africa*. Accessible at: https://climateactiontracker.org/countries/south-africa/.

Coady, D., Parry, I., Sears, L., & Shang, B. (2017, March). How large are global fossil fuel subsidies? *World Development, 91*, 11–27. https://doi.org/10.1016/j.worlddev.2016.10.004.

Cotterill, J. (2018, June 14). Power outages hit South Africa as electricity monopoly cuts supply. *Financial Times*. Accessible at: https://www.ft.com/content/63a6bd76-6ff4-11e8-852d-d8b934ff5ffa.

Davis, G. (1995). Learning to love the Dutch disease: Evidence from mineral economies. *World Development, 23*(10), 1765–1779.

DEA (Department of Environmental Affairs). (2014). *Greenhouse Gas Inventory for South Africa 2000–2010*. Pretoria: DEA.

Department of Environmental Affairs (DEA). (2018, March). *South Africa's third national communication under the United Nations framework convention on climate change*. Pretoria: DEA. Accessible at: https://unfccc.int/sites/

default/files/resource/South%20African%20TNC%20Report%20%20to%
20the%20UNFCCC_31%20Aug.pdf.

Eberhard, A. (2011). *The future of South African coal: Market, investment, and policy challenges* (Working Paper No. 100). Stanford: Stanford University Program on Energy and Sustainable Development.

EMG. (n.d.). *Water management devices: Facts and perspectives.* Cape Town: Environmental Monitoring Group. Accessible at: http://www.emg.org.za/images/downloads/water_cl_ch/FactSheetWMD.pdf.

Fine, B., & Rustomjee, Z. Z. R. (1996). *The political economy of South Africa: From Minerals Energy Complex to Industrialisation.* Boulder, CO: Westview.

Fin24. (2014, November 23). Eskom issues power emergency. *News24.* Accessible at: https://www.fin24.com/Economy/Eskom-issues-power-emergency-20141123.

Hofstatter, S. (2018). *Licence to loot: How the plunder of Eskom and other parastatals almost sank South Africa.* Cape Town: Penguin Random House South Africa.

Holland, M. (2017, March 31). Health impacts of coal fired power plants in South Africa. *Groundwork South Africa/ Health Care Without Harm.* Accessible at: https://cer.org.za/wp-content/uploads/2017/04/Annexure-Health-impacts-of-coal-fired-generation-in-South-Africa-310317.pdf.

Humphreys, M., Sachs, J. D., & Stiglitz, J. E. (Eds.). (2007). *Escaping the resource curse.* New York: Columbia University Press.

IPCC. (2018). *Intergovernmental Panel on Climate Change interim report, 2018.* Accessible at: http://www.ipcc.ch/report/sr15/.

IRENA. (2018). *Renewable power generation costs in 2017.* Abu Dhabi: International Renewable Energy Agency. Accessible at: https://www.irena.org/-/media/Files/IRENA/Agency/Publication/2018/Jan/IRENA_2017_Power_Costs_2018.pdf.

IRP. (2010, May 6). *Integrated Resource Plan, 2010–2030* (Regulation Gazette No. 9531 GG No. 34263).

IRP. (2018). *South Africa's Integrated Resource Plan.* Accessible at: http://www.energy.gov.za/IRP/irp-update-draft-report2018/IRP-Update-2018-Draft-for-Comments.pdf.

Jiborn, M., Kander, A., Kulionis, V., Nielsen, H., & Moran, D. (2018). Decoupling or delusion? Measuring emissions displacement in foreign trade. *Global Environmental Change, 49,* 27–34. https://doi.org/10.1016/j.gloenvcha.2017.12.006.

Lazard. (2018). *Levelized cost of energy data.* Accessible at: https://www.lazard.com/media/450784/lazards-levelized-cost-of-energy-version-120-vfinal.pdf.

Leger, J. P. (1991). Trends and causes of fatalities in South African mines. *Safety Science, 14*(3–4), 169–185.

Marinovich, G. (2016). *Murder at small Koppie: The real story of the Marikana massacre.* Johannesburg: Penguin.

Mearsheimer, J. (1994). The false promise of international institutions. *International Security, 19*(3), 5–49.

Meth, O. (2018). New satellite data reveals the world's largest air pollution hotspot is Mpumalanga – South Africa. *Greenpeace Africa*. Accessible at: https://www.greenpeace.org/africa/en/press/4202/new-satellite-data-reveals-the-worlds-largest-air-pollution-hotspot-is-mpumalanga-south-africa/.

Midgley, G., Chapman, R., Mukheibir, P., Tadross, M., Hewitson, B., Wand, S., et al. (2007). *Impacts, vulnerability and adaptation in key South African sectors: An input into the long-term mitigation scenarios process*. Cape Town: Energy Research Centre, University of Cape Town.

Mudavanhu, S., & Rudin, J. (2018, April 04). Renewable energy claims and counterclaims. *Daily Maverick*. Accessible at: https://www.dailymaverick.co.za/article/2018-04-04-op-ed-renewable-energy-claims-and-counterclaims/.

Nace, T. (2018, October). A coal phase-out pathway for 1.5°C: Modeling a coal power phase-out pathway for 2018–2050 at the individual plant level in support of the ipcc 1.5°C findings on coal. *CoalSwarm*. Accessible at: https://storage.googleapis.com/planet4-international-stateless/2018/10/7df76ee5-coalpathway-final.pdf.

Nakumuryango, A., & Inglesi-Lotz, R. (2016). South Africa's performance on renewable energy and its relative position against the OECD countries and the rest of Africa. *Renewable and Sustainable Energy Reviews, 56*, 999–1007.

Newell, P., & Bumpus, A. (2012). The global political ecology of the CDM. *Global Environmental Politics, 12*, 49–67.

Newell, P., & Mulvaney, D. (2013). The political economy of the 'just transition'. *The Geographical Journal*. https://doi.org/10.1111/geoj.12008.

OEC. (2018). *South Africa—Country profile*. Accessible at: https://atlas.media.mit.edu/en/profile/country/zaf/.

Sachs, J., & Warner, A. (1995). *Natural resource abundance and economic growth* (Development Discussion Paper No. 517a). Cambridge, MA: Harvard Institute for International Development.

Sasol. (2017). *Overview*. Accessible at: https://web.archive.org/web/20170831214232/http://www.sasol.com/about-sasol/strategic-business-units/energy-business/overview.

SARVA. (2018). *South African risks and Vulnerability Atlas*. Accessible at: http://sarva2.dirisa.org/.

Smith, C. (2018, March 1). Drought impact on W. Cape economy worse than anticipated—Minister. *Fin24*. Accessible at: https://www.fin24.com/Economy/drought-impact-on-w-cape-economy-worse-than-anticipated-minister-20180301.

Solomons, I. (2017, February 6). Domestic coal sales becoming more important than export market—Prevost. *Mining Weekly*. Accessible at: http://www.

miningweekly.com/article/domestic-coal-sales-becoming-more-important-than-export-market-prevost-2017-02-06.

Stoddard, E. (2017, November 7). Deaths spike in SA's deep and dangerous mines, reversing trend: Ends nine straight years of declining fatalities. *Reuters* via *Moneyweb*. Accessible at: https://www.moneyweb.co.za/news-fast-news/deaths-spike-in-sas-deep-and-dangerous-mines-reversing-trend/.

Styan, J. (2015). *Blackout: The Eskom crisis*. Cape Town: Jonathan Ball.

UN Water. (2006). *Water a Shared Responsibility*. Accessible at: http://unesdoc.unesco.org/images/0014/001454/145405e.pdf#page=519.

Walwyn, D. R., & Brent, A. C. (2015). Renewable energy gathers steam in South Africa. *Renewable and Sustainable Energy Reviews, 41*, 390–401.

WHO. (2017). *World Health Organization, Cholera data*. Accessible at: http://apps.who.int/gho/data/node.main.176?lang=en.

World Bank. (2018a). *World Bank Group data: Electricity production from renewable sources, excluding hydroelectric (kWh)*. Accessible at: https://data.worldbank.org/indicator/EG.ELC.RNWX.KH?end=2015&start=1992.

World Bank. (2018b). *World Bank Group data: Electricity production from hydro-electric sources (%)*. Accessible at: https://data.worldbank.org/indicator/EG.ELC.HYRO.ZS?locations=ZF.

World Bank. (2018c). *World Bank Group data: Coal*. Accessible at: https://data.worldbank.org/indicator/eg.elc.coal.zs?end=2015&start=1992.

Wu, G. et al. (2017). Strategic siting and regional grid interconnections key to low-carbon futures in African countries. *Proceedings of the National Academy of Sciences*, 201611845, 1–9. Accessible at: https://doi.org/10.1073/pnas.1611845114.

CHAPTER 2

Competing Paradigms for Understanding Energy Transitions

Abstract Explaining the presence or absence of energy transitions requires understanding not only a country's political history and geography, but also its place in the global political economy, and the complex interrelations of these different levels of analysis. Naturalistic accounts, emphasising the path-dependencies of natural endowments, typically neglect questions of externalities and artificially minimise political agency and power relations. Socio-technical transitions approaches allow more scope for agency, but focus principally on management strategies for niche technologies, while neglecting diverse forms of power and resistance. Critical political economy perspectives, such as those investigating the systemic nature of the "Minerals Energy Complex" and related social structures of accumulation, provide for a fuller account of local and transnational power relations, and are thus better placed to highlight counter-hegemonic sources and expressions of power. In turn, they also point to qualitatively different policy options.

Keywords Path dependency · Socio-technical transitions · Energy justice · Minerals Energy Complex · Water · Passive revolution

Explaining energy transitions—or their absence—requires understanding not only a given country's political history and geography, but also its place

© The Author(s) 2020   21
A. Lawrence, *South Africa's Energy Transition*, Progressive Energy
Policy, https://doi.org/10.1007/978-3-030-18903-7_2

in the global political economy.[1] An overly narrow focus on any one of these areas risks missing the wood for the trees. At the same time, however, by expanding a sense of transitions' complexity, scholars and policymakers broaden their scope for identifying and assessing political agency—and thereby a means of achieving not merely a transition, but rather, a transformation of economy and society, as well as energy policy. This chapter surveys a range of approaches to energy transitions that underscore several interrelated, central themes. First, such transitions are not mere technical exercises, but political projects; their potential and trajectory cannot entirely be predicted in advance. Second, the goals of energy justice and climate justice, while not identical, are interrelated in several key respects. Lastly, just as energy and climate injustices have global roots, the scope of transitions is never merely national, but includes regional and global actors, interests, and structures. Thus, however indispensable national policy change may be, it must be accompanied by local as well as global realignments and transformations.

The narrow approach is consonant with passive revolution (from "above")—the capacity of entrenched (in this instance, fossil fuel and allied) interests to deploy material, institutional and discursive power toward accommodating a modicum of pressure for more disruptive change so as to preserve prevailing social relations and distributions of political power. Mere official declarations of intent, or even specific pieces of legislation, are not only inadequate to achieve far-reaching energy transitions; they are often symptomatic of passive revolutions. This is because their implementation is mediated by incumbent economic interests whose short-term profitability is often threatened by such interventions.[2]

Just as there is no global consensus over the best means of defining, let alone achieving, a sustainable energy transition, no scholarly or policymaking consensus exists concerning the relevance of various analytical frameworks for understanding these processes. Although relatively recent, the literature on energy transitions has mushroomed to an extent that—the

---

[1] While the term "energy" is used here synonymously with familiar types (fossil fuels, nuclear, solar, wind, hydro, geothermal, and biofuels), Vaclav Smil (2006: 9–10) provides the salutary reminder that its usage is historically related to political "power": "Energy is not a single, easily definable entity, but rather an abstract collective concept, adopted by nineteenth century physicists to cover a variety of natural and anthropogenic phenomena."

[2] As Newell argues, the "power, capability and inclination among states to assume the roles ascribed to them in transition plans, their own Nationally Determined Contributions (NDCs) submitted to the climate negotiations and global sustainability strategies such as the SDGs (Sustainable Development Goals)" are clearly highly uneven. Newell (2018).

dominance of a few analytical perspectives notwithstanding—the scope and meaning of the category of "energy transitions" risks being excessively diluted and will likely remain essentially contested. At its most basic level, the term simply refers to a given polity's transition from one dominant source of energy to another, such as a transition from a dominant fossil fuel source to one or more renewable energy (RE) sources.[3] It may also connote movement from a higher to a lower energy intensity of production, irrespective of energy source. These meanings can, and often do, overlap with associated work on "sustainability transitions" that seek to transform energy and resource consumption patterns—and also, for some analyses, prevalent institutional and governance patterns—in ways that strengthen the long-term viability of ecosystems, the biosphere, and the democratic accountability and capacity of polities therein.

The "Natural" and "Objective" Constraints of Coal?

For over a century, coal has served as South Africa's major source of domestic and industrial energy use. The intensity of its exploitation, combined with the extent of its holdings and of the country's dependence on it for electricity generation, make South Africa one of the world's top coal mining and producing countries, as detailed in Chapter 1.[4] As well as providing more than 90% of its heating and electricity generation (with energy parastatal Eskom alone consuming two-thirds of total coal consumption), coal's uses include the energy- and capital-intensive process of liquefaction (or 'Coal-to-Liquids') yielding synthetic oil, petrol and diesel. This process was first promoted by the apartheid regime via its parastatal Sasol, founded in 1950 (*Suid-Afrikaanse Steenkool-, Olie- en Gasmaatskappy*, or South African Coal, Oil and Gas Company, renamed South African Synthetic Oil Company) as a means of achieving energy autonomy. Sasol consumes another 20% of annual coal use. Full energy autonomy has never been achieved, however. As shown in Chapter 1, oil products represent the biggest single import category, costing $7 billion a year. By contrast, Sasol produces via liquefaction only about 150,000 barrels per day—virtually all of the country's domestically sourced oil.[5]

---

[3] Grubler (2012).
[4] See Table 1.1, Chapter 1.
[5] Wakeford (2012).

It would be foolhardy to argue that history is destiny, and that South Africa will never be able to transition its vehicle fleet from internal combustion to electric models, for example. Yet this is a rather similar argument to what is perhaps the least convincing explanation for South Africa's tardiness in transitioning from coal dependency, one emphasizing a geostrategic, path-dependent logic to the exclusion of all other considerations. This perspective is the dominant common-sense view of many if not most leading figures in government and the mining industry. It begins with the constraints and opportunities of "natural" endowments, reads off "objective" geostrategic state interests from these, then ascribes resulting continuity to path dependency. One analysis in this vein even argued recently that "structures beyond human control"—that is, "nature"—determine

> the availability of energy resources: South Africa is rich in coal but does not possess much potential for the development of hydropower. Nature is interlinked with path dependency. … South Africa's largest power stations, which were built in the 1980s and 1990s, are all coal-fired and will continue to generate electricity for some decades yet. … The factors that explain energy policy not only comprise nature (for example, the existence of huge coal reserves in Mpumalanga) but also man-made material structures in geographical space insofar as they cannot be altered in the short run (for example, the existing structure of industrialisation in South Africa).[6]

The inexorable conclusion is that "South Africa's energy-intense economy and the predominance of coal fired power stations cannot be changed in the short run. …The most important reason for this is simply that South Africa is poor in all energy resources apart from coal."[7] The regional picture is admittedly different: "natural conditions however suggest that South African business and politicians will seek to benefit from the energy endowment of the region's countries, not only in terms of hydropower but also regarding oil and gas".[8] Ironically, the pessimism concerning domestic dependency on coal is mirrored by a naïve optimism concerning the country's continued capacity to mine and consume coal affordably, as well as the "natural" capacity to achieve a regional energy security community—overlooking the fact that worldwide, security communities in energy (or any other field) are rare, politically constructed rather than naturally occurring,

[6]Scholvin (2014: 186, 188).
[7]Scholvin (2014: 188).
[8]Scholvin (2014: 191).

and in need of constant reinforcement and renegotiation.[9] Strikingly absent from both the national and regional list of natural sources in this account, moreover, are solar and wind power, since "large-scale projects for the environmentally friendly generation of electricity do not exist."[10]

This bald statement is belied by the fact that the 2010 IRP had already planned an expansion of renewable generation of almost 18 GW, or nearly 42% of additional capacity scheduled by 2030.[11] It also overlooks the extent of RE generation already achieved, its attendant financial benefits, and thus by implication, the degree of latitude policymakers are already able to exercise in promoting RE substitutions for coal. A CSIR study calculated the net financial benefits during 2014 alone of the 2.2 TWh of wind and solar photo-voltaic (PV) generation added to the grid by that time to be R800 million, net of R4.5 billion paid to the Independent Power Producers contracted by the Department of Energy since 2011,[12] and including a valuation of avoided black-outs.[13]

These early yet positive outcomes were no anomaly, as subsequent research demonstrated. CSIR conducted a second assessment of economic benefits over the first six months of 2015. It showed that the operation of 1 GW of PV capacity delivered a net fuel cost savings of R411 million. Over the same six months, wind power delivered net savings of R389 million. In six months, 1800 MW of wind and solar PV energy provided about 2% of total supply. It replaced 500 GWh of coal-fired generation and 1500 GWh of diesel-fired "peaking plant" supply. Eskom saved R800 million; renewables contributed a total net benefit of R4 billion (or R2 per per kilowatt hour (kWh) of renewable energy) to the economy.[14] Although the draft IRP Base Case of 2016 misrepresented the lowest-cost scenario available by

---

[9] Deutsch (1961) and Adler et al. (1998); on the theory's relevance for southern Africa, see Nathan (2006).

[10] Scholvin (2014: 192).

[11] To be sure, this is an aspirational target—but can be interpreted as a plan nonetheless. Sceptical parsing shows it to be a slight exaggeration: although a 12% increase over the IRP's "Revised Balanced Scenario", the total is 33.3% (comprising 16.3% for wind, 2.1% CSP and 14.9% PV of new capacity added), or including imported hydro-electric power (4.7%), 38%, or slightly over a fifth of total capacity. See IRP (2010), Table 4, p.14.

[12] For further detail on the Renewable Energy Independent Power Producers Procurement Programme (or REIPPPP), see Chapter 5.

[13] CSIR (2015a).

[14] CSIR (2015b).

placing artificial limitations on annual additions of PV (1 GW) and onshore wind (1.6 GW) permitted, the benefits of adding RE capacity were already becoming clear by this time.

Just as the wind and solar radiation are natural endowments in the same sense as unextracted coal and oil, their use also represents an irreducibly political choice. In reference to the 2016 Mining Charter that served to enforce compliance with the 2002 Mineral and Petroleum Resources Development Act (or MPRDA, which asserts state power over the mining sector by imposing black ownership and employment requirements), industry journal *Mining Review* recently observed that uncertainty has been "the biggest cause for a decline in investments in the mining and energy sector – if the 'rules of the game' is clear – more investment flow will be seen to critical sectors of the economy".[15] This latter observation, as Chapter 5 details, also applies equally to the RE sector: it too requires strong levels of government commitment.

Nevertheless, while its fears that politics can play a decisive role in deterring (not just supporting) the coal sector are more realistic than the "naturalistic" view above, there are additional reasons for judging the journal's hopes that "clean coal" will rescue the domestic coal sector to be misplaced. Sasol alone mines over 40 million metric tons of coal annually; together with four large transnational mining conglomerates (South Africa Energy Coal—a division of Anglo American plc, South32, Glencore Xstrata, and Exxaro), it consumes over 85% of the more than 250 million tons (mt) mined annually. In terms of the diminishing quality of reserves, the detrimental health impacts already detailed in Chapter 1, negative ecological impacts, and in particular, those affecting South Africa's water supply, there are multiple grounds for doubting the short- to medium-term, let alone long-term, viability of the country's reliance upon coal.

Moreover, notwithstanding its rank among the world's leading producers and exporters, doubts remain concerning the viable extent of South Africa's coal reserves. In 2014, the government's Department of Mineral Resources estimated these at 66.7 billion tons, while observing that the rate of growth remained below 1% for over a decade.[16] This is higher than the 1987 Bredell Report estimate of 55 billion tons, of which only 2% is anthracite, and 1.6% coal of metallurgical quality. The IEA set the esti-

---

[15] *Mining Review* (2018).

[16] DMR (2014: 8).

mate even lower, at barely 30 billion tons.[17] A key distinction is between "technically recoverable" and "economically viable" reserves. As geologist Chris Hartnady observed almost a decade ago, while some reserves have a large proven size, technical features such as their structural complexity, low grade, high ash content or water scarcity are likely to inhibit major development. "Given SA's heavy dependence on coal for power generation and electricity supply, an anticipated peak production in 2020 will cause problems for future economic growth."[18] The cost of pre-combustion preparation increases as the available supply of medium- to high-grade coal (22 MJ/kg or better) has dwindled to a small fraction of the total.[19] As with oil, "peak coal" is less an issue of reserves in the ground, and more one of "peak price"—and thus "peak demand"—beyond which the economic and ecological costs become prohibitive.

Indeed, increases in the cost of coal—including an almost 25% increase from financial year 2011–2012 to 2012–2013 alone—show no sign of slackening. Eskom's coal bill was approximately R60 billion in 2018, when Eskom signed 27 new coal contracts to buy 73 mt, but this remained less than two-thirds of the 116 mt required over the short term.[20] Over the next four decades, less than half of the projected coal needs has been contracted. The fact that Eskom has become increasingly reliant upon short-term contracts of a year in duration or less, rather than "cost plus" contracts (lasting 10–40 years), is a major factor in the ballooning of costs; such short-term contracts, typically costing more, have risen from 17% of purchases in 2007 to 30% in 2012. An export price of nearly three times that of domestic consumption has also contributed to increased coal costs.[21]

However plentiful South Africa's coal may be, its water is not. At less than half a meter of average annual rainfall, South Africa's share is barely half of the global average, placing it among the thirty driest countries in the world. This rainfall level has been steadily declining over the past fifty years, and the IPCC's 4th Assessment Report expresses near certainty that climate change will continue to diminish southern Africa's water resources,[22] placing the

[17] IEA (2006).

[18] Hartnady (2010: 1).

[19] Jaglin and Dubresson (2016: 87).

[20] McKay (2018).

[21] Jaglin and Dubresson (2016: 84–86).

[22] IPCC (2007).

region's population among the two-thirds of humanity that will experience continually water-stressed conditions by 2025.[23] Already in 2000, virtually all of the country's available yield of freshwater—98%—was consumed, leaving no margin of error for increased levels of drought, pollution, or population growth.

South Africa thus has virtually no capacity to dilute pollutants and will require increasingly higher standards of water treatment that in turn will require increasingly expensive technologies.[24] Obviously, access to adequate clean water is essential for human survival, food security, and well-being in ways that coal as an energy source is not. Maximizing water's supply and quality therefore ought to be the central logic of any government policy, and indeed the 1998 National Water Act was created expressly to prioritize people's water needs and stop the extensive pollution from mine runoff that severely contaminates water supplies.

Water despoliation is but the tip of the iceberg of damage that the mining-dominated economy has caused. As Minister in the Presidency and Chairman of the National Planning Commission (NPC), Trevor Manuel, belatedly observed in an address to the National Assembly in June 2011:

> Our economic path, our settlement patterns and our infrastructure all combine to place our country on an unsustainable growth path from a resource utilisation perspective. We are the 27th largest economy in the world but we produce more carbon dioxide emissions than all but eleven countries in the world. We are a water scarce country but we use our water inefficiently. We have to change these patterns of consumption and we have to learn to use our natural resources more efficiently. We must do this with appropriate consideration for jobs, energy and food prices.[25]

These mandates are made more difficult by an absence of reliable infrastructure monitoring water supply, together with official denial of the extent of the problem. In 2013, then-President Zuma claimed that the "percentage of households with access to potable water has increased from 60% to over 90%" since 1994. However, the 2011 General Household Survey of Statistics South Africa (StatsSA) reported that only 89.5% of South African households had access to piped water—without stipulating whether it was

[23] UNDESA (2014).

[24] Turton (2009) and Wassung (2010: 7).

[25] Manuel (2011).

potable or not—of which 43.3% had piped water in their homes, 28.6% had access to water in their yards, 2.7% had access to a neighbour's tap and 14.9% had to make use of communal taps. Moreover, "access" in this context obfuscates the question of practicality and affordability; metered water, for example, may be technically "accessible" but not affordable. Since taps typically do not provide continuous access, more than a quarter of the population does not have access to a secure supply of water. The problem is worse for rural areas, where perhaps a third of the population lacks access to a reliable water supply and basic sanitation services. While rural populations suffer the most, over 26% of all schools (urban or rural), and 45% of clinics, lack such access.[26]

Available evidence suggests, moreover, that actual access to potable water is deteriorating. Given the absence of reliable and complete data, it is overly sanguine to argue, as some commentators have, that "the water sector… has performed reasonably well" at regulating itself and provided "reasonably reliable services" demonstrating that "there has been no need for an independent regulator."[27] On the contrary, the country loses over 1.5 billion cubic meters of water a year due to faulty and aging piping infrastructure. Whereas more than three-quarters of households surveyed by StatsSA rated their water services quality as "good" in 2005, by 2011 this proportion had declined to fewer than two-thirds. Only a small minority of municipalities even maintain reliable records of water quality. In 2013, the water use and waste management executive manager at South Africa's Water Research Commission (WRC), Jay Bhagwan, was quoted saying that 18% of South Africa's municipalities had poor record-keeping, 30% have "worthless records" and 13% have "no records".[28] Again, this correlates with an urban-rural as well as income divide, as the South African Department of Water Affairs and Forestry (DWAF)—responsible for water allocation—exercises virtually no monitoring of water use in rural areas.

Yet as Chapter 1 has shown, the 2018 Integrated Resource Plan (IRP) continues to maintain the centrality of coal extraction and generation. The country's 1000+ mines consume as much water as the entire population of 53 million people. Its coal mines alone are hydraulically interlinked across more than 10,000 square kilometres; gold mines cover an additional 300

[26] UN Water (2006).

[27] Muller (2013: 681).

[28] Rademeyer (2013).

square kilometres. More than 70% of the country's municipal water is drawn from surface sources which are thus especially vulnerable to mining contaminants.[29] Moreover, coal is by orders of magnitude South Africa's most water-intensive source of energy. When accounting for its extraction, processing, pollution control and disposal of by-products (but excluding water used by mine and power plant labour and water-resource restoration), coal uses almost 5% of the national water supply, rather than the frequently-quoted 2% (a figure considering only coal combustion at a power plant). Thus coal's water consumption is equivalent to between 1.534 and 3.326 l/kWh; between 5000 and 11,000 l/kWh per capita average annual consumption, or at least 350–600 million cubic meters (350–600 billion litres) annually for the country as a whole.[30] In 2017, Eskom alone consumed 307 million cubic meters (principally for cooling, and not counting water used for producing and beneficiating the coal).[31]

Among the worst and most pervasive economic and ecological costs of coal is acid mining drainage, which accelerates the leaching of toxic salts, heavy metals and radionuclides into water tables.[32] Among the worst affected water systems is the Olifants Water Management Area. Traversing the principal coal mining regions (Ermelo, Highveld and Witbank) that supply roughly half the total coal consumed, from southeastern Limpopo Province to northwestern Mpumalanga Province and the northeastern half of Gauteng Province, it provides the sole source of water upon which nearly 10 million people depend; the rate of water demand has increased over the population growth rate over the past two decades. The fact that almost half of the coal mines are open cast adds to the likelihood of water contamination.[33] The Department of Water Affairs remains negligent in exercising oversight, or cooperating and coordinating with the Department of Mineral Resources and the Department of Environmental Affairs, and allows 152 mines to abstract and discharge water without a valid licence—including 11 out of the 22 coal mines that supply Eskom.[34]

[29] Ochieng and Nkwonta (2010: 3351).

[30] Wassung (2010: 24).

[31] Gibbs and Lloyd (2018).

[32] Liefferink (2016).

[33] Ochieng and Nkwonta (2010: 3351–3353).

[34] Wassung (2010: 23).

Similarly, the Vaal River Catchment in Gauteng province—providing water to 20 million people, or 45% of the total population, as well as to firms generating most of the GDP—is at risk of suffering high levels of degradation.[35] In 2009, the Department of Water Affairs initiated the Vaal River Eastern Sub-system Augmentation Programme (VRESAP) to supply an estimated 160 million cubic meters of water from the Vaal Dam to meet the increasing water needs of both Eskom and Sasol.[36] VRESAP will thus increase the vulnerability of farmers and local communities who depend on this water for their livelihoods, especially during drought periods, and accelerate the degradation of national water supply by accommodating, rather than challenging, the coal sector and its corporate interests.

As already noted, export-quality coal reserves have the least certain supply and are increasingly costly to mine. There are additional doubts about the robustness of their intended markets. The cost per ton of South African coal for domestic consumption (lower quality and thus even more polluting) is lower (55% of sales value, but 72% of volume) than for higher quality thermal and coking coal for export (45% of value, but only 28% of volume). Under the status quo scenario, therefore, coal's longer-term future would be ever-higher volumes for export. South Asian demand grew in the first quarter of 2018 by 12%, helping push the thermal coal export total to nearly 20 mt.[37] Yet the outlook for the biggest market, India, is dim because its private investment in coal generation has ground to a halt in the face of RE price competition that is far stronger than anticipated: so much so that newly added RE capacity recently overtook new coal-fired capacity for the first time.[38] Moreover, European and Chinese demand (currently about 28% of SA exports) is waning. And even in the unlikely scenario that export demand were to continue increasing at current rates of around 6% a year—from a level currently slightly above 75 mt—it cannot hope to make up the balance of the remaining 175 mt South Africa currently consumes domestically, which (as Chapter 1 has shown) is increasingly uncompetitive with several RE sources of generation, as well as increasingly polluting.

More fundamentally, the longer-term fate of coal worldwide is uncertain at best. Whereas total investment worldwide in the coal sector was less than

[35] Van Wyk et al. (2010) and DWA (2012).
[36] DWAF (2009).
[37] SAMI (2017) and Montel (2018).
[38] Mundy (2019).

$300 billion over the past five years, for RE in aggregate this figure is $1 trillion, generating 10 million jobs.[39] The cost of onshore wind generation has fallen by around 23%—$0.06 per kilowatt hour (kWh) on average, with some schemes as low as $0.04 per KWh—and of solar PV electricity, by 73% since 2010, to $0.10 per KWh. Fossil fuels, by contrast, fall in a range of $0.05–$0.17 per KWh, and are predicted by the OECD to become uncompetitive with not only wind and solar, but a growing range of other RE sources, within five years.[40] Already, nearly half of all coal plants worldwide are losing money, a proportion that is forecast to increase to 56% by 2030.[41]

Due to shifts to services and increasing efficiency, moreover, the secular trend worldwide for decades has been that the energy intensity of production (energy consumption per unit of gross domestic product) is continually decreasing, meaning that for each unit of GDP, less energy (from whatever source) is needed. While this trend holds for OECD countries, the rate of decline is even faster among the non-OECD countries of the Global South (the overwhelming destination of exported SA coal). Since 1990, the global average energy intensity for each US dollar of GDP decreased from over 8 thousand British thermal units (BTUs) to about 5.5 thousand.[42] Therefore, ever-stronger price competition from cleaner sources is coal's fate in the immediate term, and the opportunity costs for delaying a transition from coal grow exponentially in the longer term.

Quite apart from overlooking the costs of coal and conversely understating the actual gains and real potential of RE sources, the "naturalistic" approach to South Africa's energy transition serves to naturalize politics and thereby reduce if not entirely overlook the true scope of political agency in crafting energy transitions. In a sense, this bias is akin to the patriarchy and white supremacy intrinsic to the colonial and apartheid eras, which used ideologies of race and gender to naturalize politics. For the same reasons, it also tends to occlude the identities and networks of vested interests engaged in preserving the status quo.

---

[39] IEA (2017).
[40] IRENA (2018).
[41] Carr (2018).
[42] EIA (2016).

Socio-Technical Transitions: RE Transitions as Critical Conjuncture

The socio-technical literature on energy transitions (or STT) explores the interaction of elements of a "socio-technical system"—the complex inter-relation of a workplace, firm, or broader society and its dominant tech-nologies—across several levels, to understand how transitions could come about.[43] Work in this area—often funded by, or written from the per-spective of, major RE industry players—typically adopts three analytical levels: *niches* (the locus of radical innovations), sociotechnical *regimes* (the locus of established practices and associated rules that enable and constrain incumbent actors in relation to existing systems), and an exogenous socio-technical *landscape*.

The (typically goal-oriented) analysis progresses up to a series of "land-scape" pressures, including climate change and shifts in international energy markets which exert disruptive (principally economic) pressure upon a pre-vailing energy system regime, that may ultimately facilitate a transition away from the dominant status quo ante. Change in this perspective is con-ceived as resulting from pressures from above and below that can "lead to cracks, tensions and windows of opportunity."[44] This agent-centred focus highlights actors and pathways that can bring about transitions in energy systems. While not intrinsic to this approach, the STT literature is over-whelmingly preoccupied with national-level transitions.

As an "ideal-typical" abstraction of real-world cases, the multi-level framework of STT analyses necessarily homogenize actors, identities, and roles; the analytical distinctions among levels are also inevitably arbitrary.[45] The logic that this literature posits regarding RE's disruptive potential for energy markets is, however, intuitively plausible. While disruption can derive from a variety of sources, including energy shortages, regime change, political conflict, and changing cultural norms, with specific reference to renewable energy, it entails a near-simultaneous change in how electric grids are organized, how electricity is generated and stored, and how it is consumed.

---

[43] Geels (2005) and Loorbach (2007).

[44] Geels (2010: 495).

[45] Smith et al. (2005).

The typical pattern for twentieth century electricity industries, in South Africa and elsewhere, entailed a "merit order" (or logical sequencing of power generating sources) that placed baseload facilities that must operate constantly to meet off-peak (or "shoulder") demand first; then would secondarily draw upon stored raw energy (such as natural gas and diesel, or pumped water storage) to quickly generate electricity and thereby meet momentary increased levels of peak demand. Total operation costs were met by charging a premium for peak demand, by setting tariffs far above their marginal cost, or both. The system presumes passive consumers and offers little incentive to increase efficiency, and less in the way of oversight or participatory governance.

Even if coal were not the principal energy and electricity generation source, this baseload model is costly and inefficient—even more so for large countries with widely dispersed grids such as South Africa. Over the past few decades, this structure has become increasingly expensive in terms of input costs and inefficiencies (deriving from increasing fossil fuel volatility, wastage from long-distance generation, and blackouts), as well as social costs of pollution, emissions, and more capital-intensive infrastructure. Mega projects with typically ballooning budgets prone to costly delays and lock-in effects—such as South Africa's Medupi and Kusile coal plants and its Koeberg nuclear power plant (discussed in Chapter 4)—are characteristic of this model.

The emerging twenty-first century alternative is a decentralized model that uses "smart grid" technology to integrate smaller-scale distributed generation with actively managed demand.[46] Dispatchable RE sources, with typically much lower (indeed, close to zero) operating costs, serve to flatten peak prices. The old merit-order favouring 24/7 baseload supply running as close to 100% (or maximum rated) capacity as possible is inverted, since (as was shown in Chapter 1), wind and PV—although not continuous—are already 40% cheaper than new coal plants. They are also much more modularly scalable: In general, the smaller the generation inputs, the shorter lead times they need to be built and added to the grid, the lower the risk of cost overruns, and the easier they are to move on and off the grid as fluctuating demand dictates. As decentralization proceeds across the grid, the argument that wind and solar generation are incapable of providing secure and stable power (since "the wind doesn't always blow and the sun doesn't

---

[46]Wilson et al. (2015).

always shine") is increasingly shown to be inadequate, if not spurious.[47] A UBS report from five years ago already predicted that "large-scale power generation will be the dinosaur of the future energy system: Too big, too inflexible, not even relevant for backup power in the long run."[48] This alternative represents by no means an inevitable choice at present, however, but should rather be understood as a conjunctural opportunity: the longer the transition to it is delayed, and the more slowly it is completed, the more the opportunity costs for doing so grow exponentially.

While useful in highlighting the *potential* disruptive force of emergent or so-called "niche" technologies (such as wind and solar) in policy regimes, the socio-technical transitions literature typically neglects questions of politics and power—and related ones of timing—beyond specific management strategies and governance practices.[49] Niches are not actors and are thus incapable of opening "windows" of opportunity, which are exogenous to niches and derive from systemic effects. Geels, the *eminence grise* of the literature, admitted as much in more recently concluding that STT literature needed to pay more attention to regime dynamics and more particularly to "instrumental, discursive, material and institutional forms of power and resistance" in order to "better understand not just regime resistance but also the destabilization and decline of existing regimes, i.e. the 'destruction' part of Schumpeter's 'creative destruction' concept".[50]

In much of this literature, moreover, there remains a reductive focus on decarbonisation—the reduction of greenhouse gas (GHG) emissions to levels required to keep global temperature rise under a level of 2°C or even 1.5°C—rather than a broader approach encompassing political requisites and consequences of transitions.[51] Such reductions admittedly would entail a significant, often radical, reordering of most regional and national economies and environments, especially those in the OECD and China, where aggregate and per capita emissions levels are highest. Even so, there are a number of dangers inherent to this reductive vision. It tends to emphasize putatively technocratic solutions (such as carbon taxes) over

---

[47] South Africa already makes use of concentrated solar power, as well as pumped hydro-electric power—a mode of storage that can be powered by sources including wind and solar. These can be made universally available at a net savings estimated at millions of lives and thousands of US dollars per capita per year worldwide: Jacobson et al. (2017).

[48] UBS (2013: 1).

[49] Shove and Walker (2007) and Scarse and Smith (2009).

[50] Geels (2014: 28, 37).

[51] E.g. Geels (2018).

explicitly political tradeoffs (such as the need for divesting from fossil fuels). It shares an elective affinity with technologies (such as carbon capture and storage) whose risks and efficacy are not fully known or proven. It also subordinates questions of policy process—ones underscoring the importance of broad-based consultation, deliberation, participation, and empowerment of affected communities—to end results. Technical sustainability can thereby become divorced from political sustainability.

As suggested in Chapter 1, for countries like South Africa outside of the OECD and China, such considerations of process and participation are arguably more decisive, since their aggregate emissions are not as high. Their political regimes are also often weaker and societies are less resilient in the face of increasingly rapid climate change. The foregoing discussion of coal's use and despoliation of water suggests just one aspect of the mutual constitution of political and ecological sustainability. A reductive emissions-centred focus also often flattens geographical specificity and foreshortens historical context. The diversity of relevant scholarly fields (engineering, politics, geography, economics, etc.) alone accentuates a further point, as much political as it is academic: it would cast doubt on claims from any one of these—or from *soi disant* sociotechnical transition experts—to claim a monopoly on relevant energy-related knowledge.

JUST TRANSITIONS (JT), ECOLOGICAL, AND CRITICAL POLITICAL ECONOMY PERSPECTIVES

Too often, analysts and policy makers presume—rather than demonstrate—a zero-sum trade-off among the goals of reducing emissions and pollution, increasing energy access and affordability to communities suffering from legacies of marginalization and oppression, and improving security of (national, regional, or local) supply. This trade-off is commonly termed an "energy trilemma" among the contending goals of attaining clean, affordable, and secure energy supplies.[52] As such, the implicit framework is one of neoclassical economics' presumption of scarcity.

There are several reasons for scepticism about the factual basis for this dilemma which in turn raise questions about how economic and normatively qualitative criteria relate. As already noted, RE sources are not only the cleanest, but also, increasingly, the most affordable. In addition, by

---

[52] Among the earliest references to the concept of energy trilemmas is found in work advocating for nuclear power: see, e.g., Starr (1999).

stabilizing the grid through its increased decentralization and diversification of energy types, energy security may also be strengthened; in short, the dynamic may be "win-win" rather than "zero-sum". Since decentralization also requires more active and localized participation, the domestic rationale for RE transitions extends beyond RE cost competitiveness and ecological effects, to include the creation of feedback loops that more substantially promote participatory governance; policymakers can thereby better calculate demand and externalities.

The question is thus not only one of which technologies a country should adopt as its preferred incremental resource, but more fundamentally, one of its electricity system's fundamental structure, organizing principles, and the political cultures upon which the system most depends. A progressively decentralized grid increasingly requires more delegated responsibility and governance transparency, both of which are typically requisites and products of increased political trust. Yet the mere introduction of smart grid technology does not in itself guarantee that local communities' access to energy or scope of decision-making will be improved. Nor would the universal introduction of net metering—which allows households and firms to sell locally generated electricity back to the grid—on its own ensure that currently underserved, poorer and more marginal communities with long histories of suffering energy injustice would thereby benefit.[53] Nonetheless, the implications of such a change are potentially transformative.

The STT literature's deficit regarding more holistic accounts of sustainability is better addressed by "just transition" (JT) scholarship and policy, infused with normative concern for past and present energy injustices.[54] This literature seeks to elucidate the relationship between development processes and energy transitions—both in South Africa and with comparative reference to the global South—over the past decade, at both global and local levels. Some scholars highlight the significance of the UN's Sustainable Development Goals (SDGs) and Future Earth partnership between the natural and social sciences as signalling the interconnectedness of the previously largely separate policy goals of human well-being and environmental sustainability. They seek to determine whether and under which conditions the twin dynamics of development and energy transitions arise in parallel, in tandem, or in contradiction to one another in the context of national-level policy-making.

[53] Monyei et al. (2019).
[54] Swilling and Annecke (2012).

Just as the socio-technical approaches improve upon naturalist accounts in attempting to specify actors, and conditions under which, energy transitions take place, so the JT scholarship adds contextual perspectives on policy formulation and implementation dynamics of state-led development and sustainability transitions. In particular, Swilling, Musango and Wakeford extend the STT literature's notion of socio-political regime to encompass a space of policy related action and engagement by a wide range of actors within and outside the formal political system operating in four dimensions: power dynamics, paradigm commitments, state organization and policy programmes. For a "just transition" to succeed, they argue, the creation or further development of "publicly accountable institutions" that can exercise relative autonomy from neoliberal governance norms is necessary.[55] More than that, however, they underscore the necessity, however politically unlikely at present, of pursuing this development through a commitment to the well-being of South Africa's communities and ecosystems.[56]

This emphasis builds on earlier work on state autonomy and capacity that advocated structural transformations from exclusive coalitions focused on short-term credit- and consumption-led growth (especially in such sectors as finance, insurance, real estate, communication, and catering sectors) toward inclusive coalitions focused on satisfying basic human needs in production-focused sectors (such as agriculture, water, electricity, light manufacturing, and construction).[57] In some key respects, however, it contrasts with the typical emphasis of the older developmental state literature, particularly on the questions of the scope and type of further industrialization, and of civil society or popular participation and mobilization. While both agree on the need for deep structural transformation of the economy as well as supportive state institutions and practices, the goal hypothesized in much developmentalist literature is "accelerated economic development that substantially raises the average GDP per capita with a focus on industrialization and urbanization", whereas for the just transitions literature envisions "a low-carbon resource-efficient economy" achieved through state coordination of "a wider set of class alliances" via "a multiplicity of smallish interventions, rather than a few, massive, physical infrastructure invest-

---

[55]Swilling et al. (2015: 2).

[56]Swilling et al. (2015: 17–19).

[57]Edigheji (2010), Evans (2010), Kohli (2006), and Lawrence (2013).

ments that satisfy the need for capital deepening, but do little to redefine the institutional context for the circulation of the benefits."[58]

In this view, unless underlying power dynamics and paradigm commitments (entailing not just the definition of policy challenges, but the beliefs surrounding them) are transformed, just transitions are unlikely to succeed.[59] An open question (addressed in the concluding chapter) remains how "large"—extensive, expensive, and/or long-lasting—state-coordinated interventions would prove to be necessary in order to address not only global climate concerns, but also older, and equally pressing, more local demands for energy justice: that is, for access to adequate, clean, affordable, and secure supplies of energy for communities that have never experienced them before. Nevertheless, a central contention of the JT literature (shared by this book) is that "sustainability transitions" are entirely feasible for society to provide. Implicitly and more radically, however, the challenges of food and water insecurity, global warming hazards, and financial crises are not subordinate to economic policy, in this view, but rather define the contours to which it must adjust for the well-being of all.

Although the demand for energy justice remains pertinent among marginalised communities in most countries worldwide, it is especially so for large proportions of virtually all middle- and low-income countries. At least 16% of the global population—more than 1.2 billion people—lack access to secure and adequate electricity supply; the overwhelming majority—80%—live in rural areas, above all, in sub-Saharan Africa. Despite the shortcomings of achieving universal access, the importance of community access to electrification for well-being are widely recognized, in terms of improved health and education infrastructures, food security, and clean and safe home and community environments.[60]

In the case of South Africa, the destructive effects of three centuries of colonialism—resulting from its minority settler population's progressive dispossession of and subsequent denial of access to land previously occupied by the country's indigenous majority, together with restrictions on freedom of work, ownership, settlement and movement—were already apparent by the end of World War II. These were exacerbated by more than four decades of apartheid rule, which intensified and added to the colonial

---

[58] Swilling et al. (2015: 3, 5).

[59] Swilling et al. (2015: 8–9).

[60] Lawrence (2018).

and Union era's laws, institutions, and practices of racial domination and exclusion. As with other spheres of economy and society, energy policies also betrayed heightened effects of inequality and exclusion, from denying blacks equal and adequate access to the electrical grid, to enforcing worker exploitation, environmental racism, and political repression in the service of South Africa's corporate mining interests.

The legacies of apartheid policy have not been eradicated—or even, in some cases, substantially mitigated—after almost a quarter century of post-apartheid democracy. For example, despite having one of the lowest percentages of the population living in rural areas (around one third) among sub-Saharan African countries, and despite the National Electrification Programme (NEP)'s goal of delivering 100% access of subsidized electricity to low-income households by 2003, at least 14% of South Africans currently lack adequate energy access—a proportion that has grown since 2014.[61] In 1999, a majority of rural households still had no electricity, and by 2017, this remained true for at least 25%; a third of the population lack electricity for cooking, and off-grid electrification remains negligible.

This is shown by the extent of participation in the government's Free Basic Alternative Energy Policy, which instructs municipalities to select suitable off-grid energy sources – including paraffin, liquid petroleum gas, bio-ethanol gel (or fire gel), and coal—and make them available according to means-tested criteria to their poorest households. Although the policy's reach is well short of the entire eligible population, nearly a quarter of South Africa's local and metropolitan municipalities—49 out of 213, clustered in the poorest and most rural parts of the country—provide poor households with at least one form of off-grid energy source, according to the 2017 non-financial census of municipalities report.[62]

Most of these are inflammable and thus more dangerous than electricity. For example, more than 86,500 poor households in 20 municipalities (in the Eastern Cape, Northern Cape, and North West provinces—2.5% of the 3.5 million poor households nation-wide) receive free paraffin (i.e. kerosene, used for heating and cooking in almost 15% of all households). More than 2.5 million households (5% of all households) use candles as a main source of lighting, but fewer than 13,700 poor households in seven municipalities (in the Eastern Cape, Northern Cape, and North West

[61] World Bank (2019).
[62] StatsSA (2017).

provinces) enjoy free municipal provision. Fewer than 20,000 households in ten municipalities (in the Eastern Cape and KwaZulu-Natal) receive free fire gel, a cleaner cooking alternative to most solid fuels.[63] Again, this is far short of total need, since household air pollution from fuels such as wood, coal and dung (which are used in more households than kerosene) affects over 6.6 million people in more than 1.4 million households, including more than 41% of rural households, leading to nearly 8000 deaths a year.[64]

For many in urban areas as well, access is inadequate, despite receiving disproportionate provision of free solar electricity home systems (or SHSs: over half of the 113,000 households in 22 municipalities receiving them are in Johannesburg, Tshwane and Ekurhuleni, the three major metropolitan municipalities of Gauteng). The typical SHS provides only 50Wp, a level inadequate for cooking, colour television, or refrigeration, albeit more than the NEP's Free Basic Electricity allocation of 50 kWh/month to poor households that is meant to satisfy (but falls far short of) universal basic needs. Additionally, for the urban majority without an SHS, the metering infrastructure often malfunctions and increases overall household indebtedness. It also does not extend to all urban communities. Informal housing built on land not approved for electrification cannot get a metered connection, with a subset relying on extension cords to neighbouring metered houses at twice the cost. The poorest communities thus lack adequate access and continue to suffer from apartheid geography's legacies.[65]

More fundamentally, such energy injustices are both connected to and symptomatic of the broader structures of poverty, inequality, environmental injustices, and social and political marginalization. Regarding inequality alone, there is no improvement from the apartheid to the post-apartheid eras. In 2005, outgoing Eskom Chairman Reuel Khoza boasted that

> the 2004 celebrations of 10 years of democracy were of great significance. Ten years earlier South Africa had ... a government intent on being exclusionist, a polarised society, internal conflict, a global resistance struggle and an ailing economy. The political miracle of 1994 led to the first democratic elections and the thrust towards reconstruction and development. Ten years later, the country has a new political order, peace, rapid economic growth

[63] General Household Survey (GHS) 2016 report (GHS-SA 2016).

[64] Clean Cooking Alliance (2018).

[65] Martinez and Ebenhack (2008).

and impressive gains in areas under government control such as fiscal, monetary, trade and industrial policy. South Africa today boasts the highest level of macroeconomic stability in 40 years. There has also been great progress in addressing poverty and inequality.[66]

There is little reflection here on whether a reductive focus on macroeconomic stability of four decades prior was related to its undoing—or could become so again. Regardless, the assertion of "great progress" in addressing inequality is belied by the facts. At the height of apartheid, in 1970, the richest 20% of the population owned an estimated 75% of all wealth; by 2011, the top 10% earned almost 60% of all income, and owned approximately 95% of all wealth; the bottom 80% own no wealth.[67] These structural factors call into question not only current energy policies, but also those pertaining to (national and transnational) investment and profit flows, related epistemological politics, and the adequacy of current responses across several political and geographic scales that fail provide recognition of these harms or express solidarity in attempting to rectify them.[68]

Perhaps the most prominent of these harms is that of mass unemployment, which has been a feature of the South African political economy for decades, and has gotten worse since the global financial crisis. In 2017, 9.3 million working-age South Africans—almost 37%—were unemployed or discouraged from seeking employment, a rate that worsened in 2018 to 40%. The National Planning Commission's "National Development Plan 2030" states a goal of reducing unemployment to 6%, but this would entail either historically unprecedented levels of sustained per capita GDP growth (over 5%) continuously over almost two decades, or the state provision of over 11 million jobs—a million permanent positions per year for over a decade.[69]

South Africa presently has little prospect of sustained levels of sufficient growth; the GDP growth rate has remained below 4% for over a decade.[70] Annual real GDP per capita growth plummeted from 1.8% in 2008 to − 2.7% in 2009 following the global financial crisis. Since 2011, it has con-

---

[66] Eskom (2005).

[67] Webster (2017).

[68] Bickerstaff et al. (2013).

[69] NPC (2012).

[70] Trading Economics (2018).

sistently declined, contracting by 1.1% in 2016.[71] Even if there were such a prospect, it is unlikely on its own to significantly reduce structural unemployment, because the benefits of growth accrue most to countries with the lowest levels of inequality, and least to those (such as South Africa, which remains among the most unequal countries in the world) with the highest levels. Countries with very low Gini coefficients (measuring overall inequality) have achieved reductions in poverty of 4% for every 1% increase in annual per capita income, whereas those with high Gini coefficients (45 or higher: virtually all countries in central and southern Africa, Central and South America, and several in Asia, with populations totalling over 2 billion) typically achieve less than 1% poverty reduction.[72] Conversely, however, reducing unemployment has a positive, direct, short-term effect on growth: over one percentage per year for every 5% unemployment reduction, by some estimates.[73]

There are several reasons why labour-intensive policies leading to immediate decreases in unemployment may have a significant and positive knock-on effect on growth. A Keynesian perspective on labour markets well understands that labour markets do not clear in the way that neoclassical models posit and hence there is involuntary unemployment. But post-Keynesian perspectives relevant to South Africa's massive levels of structural unemployment, following Kalecki's model of "wage-led economic growth," add the additional insight that since firms have unused capacity and workers have a higher propensity to spend their income than do capitalists, an increase in labour's share of income boosts aggregate demand—to which firms immediately respond by increasing output rather than prices. Once incentivized to invest, they bring capacity utilization back towards its normal level. In other words, there is a positive relationship between the wage share and output, with increasing wage share leading not only to higher consumer demand, but also to higher investment and capital stock expansion.[74]

This goal is all the more urgent in the particularly dire case of youth unemployment: out of 10.3 million South Africans between the ages of 15 and 24, a large majority was unemployed, one of the worst

---

[71] DPRU (2018: 2).

[72] Ravallion (2007: 37–61).

[73] SAMI (2018).

[74] Stockhammer (2015).

rates of youth unemployment worldwide. Even worse, most these unemployed youth—over 3 million—are categorized as NEET: "not in employment, education, or training".[75] Since wages from work remain by far the most important source of income for poor household members, the consequences of a vicious cycle of structural unemployment are devastating. For example, childhood stunting and malnourishment which permanently damage physical and mental development, are alarmingly widespread: the 2016 South Africa Demographic and Health Survey found that 27% of all children under five suffer from stunting; other studies suggest even higher rates.[76] Those with the least responsibility for the injustices and policy failures of apartheid and the post-apartheid status quo are among those bearing their brunt most heavily, lacking access to—or indeed any prospect of—decent work as a path out of poverty.

To a significant extent, chronic unemployment, especially among less skilled workers, can be successfully addressed through clean energy transitions. Because RE jobs have a higher level of labour intensity, inward investments in wind and solar, for example, generate more jobs on average than more capital-intensive coal, gas, and nuclear plants do. More broadly, when spending on RE investment and efficiency improvements to infrastructure (such as on public transport, retrofitting buildings, and producing and installing solar panels) stays within the domestic economy rather than expended on imports, the multiplier effect further increases GDP growth. Moreover, while technological improvements typically increase average labour productivity, growth in operational activities and investment spending typically increase at a higher rate, meaning that employment expands with increased standards of living, particularly in agriculture (if bioenergy is pursued), manufacturing, and construction.[77] Rather than requiring an entirely new set of skills, such jobs typically need merely an upgrading of existing skills. A major advantage of RE jobs in this regard is that they can be tailored to specific regions—such as those currently most affected by coal mining, and those with the highest levels of chronic unemployment and underemployment.[78] Aggregate reductions in unem-

---

[75]DPRU (2018: 14).

[76]SADHS (2017) and Steenkamp et al. (2016).

[77]Promoting a sugar and bioenergy industry in South Africa alone could generate over 100,000 jobs annually, especially for depressed rural areas of KwaZulu Natal and the Eastern Cape. See Ward (2018).

[78]Pollin (2015: 80–82).

ployment—which remain one of the best ways to stimulate growth—are also the best overall guarantor of worker empowerment and wages over time.

A key insight that much of the JT and critical political economy scholarship share concerns the organization of South Africa's dominant sectors as a Minerals Energy Complex (MEC).[79] This analytical perspective understands the South African economy as having evolved as a public-private partnership on a grand scale organized around the mutually reinforcing mining and energy sectors. The development of parastatals such as Eskom ensues from this partnership as a means of accommodating political interest group challengers to dominant corporate and financial interests, cemented by a common concurrence over the practical development of apartheid. While the first phase of "grand apartheid" roughly coincided with the attraction of foreign direct investment, the rise of monetarism ushered by the collapse of the Bretton Woods system and collapse of the gold standard intensified the state's strategy of import-substitution industrialization, using income from gold mining for inward (and increasingly militarized) investment.[80] This perspective is largely congruent with Social Structures of Accumulation (SSA) theory's analysis of long term trajectories of growth and decline. Developed initially to explain the decline of most OECD economies in the wake of the collapse of the Bretton Woods System and global economic slowdown of the early 1970s, this approach focuses on the ways in which specific SSAs—assemblages of economic, political and ideological institutions—promote distinct trajectories of capitalist accumulation, first as enablers of, then as fetters upon, economic growth.[81]

Monetarism also heralded, particularly from the 1990s onwards (worldwide no less than in South Africa), the rise of finance to the detriment of real capital formation, privatisation, outsourcing, and the concomitant trumping of shareholder value over all other social goals, with the accumulated effect of eroding state capacity. It was also to the detriment of investment (particularly in the manufacturing sector), and thus by extension to real productivity growth, wages, and employment levels. The moment when the 2012 National Planning Commission identified its regime as a "developmental state" ironically underscored the extent to which it fell short

---

[79]Fine and Rustomjee (1996) and Mohamed (2010).

[80]Lawrence (2014: 247–249).

[81]Weisskopf (1979) and Lippit (2010).

of the "East Asian Tigers" model of state-led development.[82] Financialisation also facilitated the means of accommodating the aspirations of the newly enfranchised black majority—via Black Economic Empowerment (BEE) policy's enrichment of a politically connected minority—echoing the interwar pattern of accommodating political interest group challengers without substantially challenging the MEC's structures of accumulation: a classic instance of passive revolution. At the same time, it has also greatly accelerated capital flight, to levels approaching 20% of GDP in 2007: "a consequence of transfer pricing by corporations, specifically in and around mining." Moreover, "the South African Reserve Bank's 'Voluntary Disclosure Programme' effectively amounts to an amnesty for capital flight and, given that the proposed 'amnesty' is part of an ongoing process moving towards the complete removal of exchange controls, the repatriation of wealth [out of the country] is likely to increase in the future rather than to fall."[83]

Following Kalecki and others in the post-Keynesian tradition, the SSA perspective understands institutional legitimacy as a key determinant of the durability of regimes of accumulation—and by extension, of the speed with which they can disintegrate.[84] South Africa's transition from high levels of growth in the early apartheid phase of the 1940s–1960s (fuelled by the forced repression of black labour costs, infrastructure development, and momentary macroeconomic stability), to stagnation and crisis from the late 1970s to the present, is mirrored by the nonfinancial corporate sector's profit rate shifting from fluctuations between 20 and 35% in the 1960s and 1970s to a level rarely exceeding 10% from the 1990s to the present.[85] It is also predictable that BEE policy would be adopted at this latter period, precisely because it serves as a distraction from the failure of Growth, Employment, and Reconstruction (GEAR) to achieve sustained high levels of growth.

Understanding the nature of South Africa's democratization as an "elite transition" along these lines also improves upon the STT scholarship in drawing concrete connections among domestic and international

[82] Ashman et al. (2010).
[83] Ashman et al. (2011).
[84] Kalecki (1971).
[85] Heintz (2010: 273).

constraints and opportunities.[86] For example, in response to the wave of blackouts and load-shedding in 2008 (detailed in Chapter 3), the South African government announced revised plans to construct the world's largest two coal-fired power stations (of about 4800 MW each) with a planned operating lifespan of 50 years, Medupi and Kusile—despite international and local civil society condemnation of a projected annual level of emissions of 25 mt of CO_2 (together responsible for a majority of the increased emissions projected between 2020 and 2040; this figure has since been revised upward to 32 mt CO_2).[87]

What bears emphasizing from a critical political economy perspective is that their construction on the proposed scale would not have begun without crucial international financing, provided by actors and institutions that pay lip service to decarbonisation goals. The Medupi Plant's $3.75 billion World Bank loan (at the time, the Bank's largest ever) was endorsed by then-U.S. Secretary of State Hillary Clinton, notwithstanding campaign positions on climate change and official U.S. abstention when the loan was voted upon by the Bank.[88] Resource curse effects became immediately apparent: in 2009, the South African public protector found Medupi's financing entailed "improper" conflicts of interest with the government's ruling African National Congress (ANC), in power since 1994. Then-chair of Eskom, Mohammed Valli Moosa, also an ANC Finance Committee member at the time, approved Hitachi as the main supplier of the plant's $3.3 billion boilers, while aware that the ANC's Chancellor House had a 25% share of its local subsidiary and thus would stand to make approximately R1 billion from the deal.[89] Meanwhile, the Hitachi boilers and auxiliary equipment, as well as turbines provided by Alstrom A&E, suffer from major technical failings that have been a major cause of delays in the plant's inauguration.[90] As Bobby Peek, director of the community development NGO Groundwork argued at the time, "This loan will put South Africa deep into debt, damage the environment and drive the

---

[86] Bond (2000, chapter 1).

[87] Sourcewatch (2018a, b).

[88] Vardi (2016). While campaigning for US president, Clinton promised that "as President, I will work both domestically and internationally to ensure that we build on recent progress and continue to slash greenhouse gas pollution over the coming years as the science clearly tells us we must" (Harrington 2016).

[89] *Mail & Guardian* (2010).

[90] Yelland (2015).

climate impacts already affecting poor South Africans. It is not electricity for the millions of people who live in deep rural areas who still have no electricity. It's for big industry which uses more than 80% of South African electricity."[91]

The other plant, Kusile, received $850 million in funding from the U.S. Export-Import (Ex-Im) Bank, a US government agency led by Fred Hochberg, a close Clinton ally and fundraiser. The plant is being built by Black & Veatch, a U.S. construction firm tied to another Clinton associate, H.P. Goldfield (also a former Ex-Im director and vice-chair of the lobbying firm of Madeline Albright, President Bill Clinton's Secretary of State). In Wikileaks-leaked diplomatic cables written in the latter part of 2009 and early 2010, Ambassador Donald Gips provided the State Department with background information on the energy project and outlined some of its related major environmental and economic concerns.[92]

It is noteworthy that the Gips memo promises that while the two mega-plants, together with "previously moth-balled coal-fired stations" will satisfy short-term needs and will be commissioned between 2012 and 2013, "in the longer term, Renewable Energy Feed-in Tariff programs (REFIT), Medium Term Power Purchase programs (MTPPP), and the open-cycle gas turbine (OCGT) independent power producer deals (IPP) are expected to augment capacity in the medium term. On the demand reduction side, Eskom's IRP includes … demand side management projects such as the installation of one million solar water heaters, a solar concentration facility, a nuclear fleet strategy (to provide low emission base-load alternatives to coal-fired generation from 2020), hydro capacity from the region, and a gas-fired option at Moamba, Mozambique." Regarding nuclear power, it adds that South Africa "remains committed to expanding nuclear power and Westinghouse remains well positioned when the process and timing are clarified, expected also in early 2010."[93]

The memo provides no justification for its proposed sequencing; it is doubtful that any plausible one in South Africa's interest could be given. Rather, the its motivation appears to be purely one of coal and international investor profitability. The huge sunk costs projected for the Medupi and

---

[91] Brkic (2010).

[92] WikiLeaks (2010).

[93] The REFIT program is discussed further in Chapter 3; the nuclear power construction plans, in Chapter 4; and the IPP programme, in Chapter 5.

Kusile plants not only cast the government's emissions reduction commitments into doubt, but starkly illustrate how (especially large-scale) fossil-fuel and nuclear generation still enjoy an advantage in terms of government prioritization and access to (especially international) financing, compared to RE generation.

In virtually every instance, RE infrastructure and demand reduction proposals could be achieved sooner and at lower cost that the mega-plants' construction. The contract for the Sere Farm installation was awarded in 2013; costing R2.4 billion, Sere began feeding electricity into the grid after one year, and became fully operational within two years of award of contract.[94] Although promising to deliver 48 times more electricity, by the time the Sere contract was awarded, *Business Day* already had Medupi's total construction cost (not including externalities) estimated at a minimum of R150 billion (more than 50 times Sere Farm's cost), the most expensive of its kind in the world, with full costs potentially exceeding R1 trillion.[95] In 2010, public enterprises minister Barbara Hogan defended Medupi by saying "there is no Plan B. … If we do not have that power in our system, then we can say goodbye to our economy and to our country. This is how serious this thing is. The construction of Medupi … is necessary so that we do not derail the country's economic growth and development."[96] It should have been clear by this time, however, that large power plants were especially vulnerable to cost overruns and delays. The first unit did not become operational until 2015. By the end of 2018, Medupi was still not fully operational.

The fact that global finance remains a dominant actor in resisting and undermining renewable energy transitions is shown by recent trends in accelerated investment into fossil fuel production. Over a dozen transnational financial concerns—each holding over a trillion dollars in assets under management, together holding over forty trillion—in 2017–2018 increased their holdings in fossil fuels. The two largest—BlackRock and Vanguard—together hold over 11 trillion in assets under management, including roughly 4 billion tons of thermal coal, which if consumed would emit over 8 gigatons (Gt) of CO_2e alone. Unlike oil and gas, however, public companies (themselves substantially owned by pension and policy

---

[94] Creamer (2013).

[95] Gleason (2013).

[96] Quoted in Brkic (2010).

holders, government funds, endowments, etc.), hold approximately half of all thermal coal assets, thus increasing exposure to perceived public risk as well as market pressure.[97]

Conclusion

The contrasting perspectives on South Africa's energy transitions—with comparative relevance to other fossil-fuel dominant countries—do not simply highlight various facets of a large and complex topic. They point to substantially divergent policy solutions—a point to which we return in the concluding chapter. The "naturalistic" approach—exemplified by official justifications of coal generation for providing "indispensable baseload power"—has nothing to say about externalities, transnational influences, or indeed suboptimal technical and economic performance, let alone the connections among these factors. Improvements are to be sought exclusively through South African exports' enhanced competitiveness and, when necessary, increased austerity at home. The STT approach informs and applauds the REIPPPP programme, but similarly subordinates (or ignores) energy injustice and economic transformation issues, particularly surrounding employment creation. The latter are, by contrast, a central focus of the JT literature, critical political economy perspectives, and more important, the grassroots campaigns informed by these perspectives—discussed in greater detail in the conclusion.

References

Adler, E., Barnett, M., & Smith, S. (Eds.). (1998). *Security communities*. New York: Cambridge University Press.

Ashman, S., Fine, B., & Newman, S. (2010). The crisis in South Africa: Neoliberalism, financialization and uneven and combined development. *Socialist Register, 47*, 174–195.

Ashman, S., Fine, B., & Newman, S. (2011). Amnesty International? The nature, scale and impact of capital flight from South Africa. *Journal of Southern African Studies, 37*(1), 7–25. https://doi.org/10.1080/03057070.2011.555155.

Bickerstaff, K., Walker, G., & Bulkeley, H. (Eds.). (2013). *Energy justice in a changing climate: Social equity and low-carbon energy*. New York: Zed Books.

---

[97]InfluenceMap (2018).

Bond, P. (2000). *Elite transition: From apartheid to neoliberalism in South Africa*. London: Pluto Press.

Brkic, B. (2010, April 7). High noon for Eskom's World Bank loan bid. *Daily Maverick*. Accessible at: https://www.dailymaverick.co.za/article/2010-04-07-high-noon-for-eskoms-world-bank-loan-bid/.

Carr, M. (2018, November 30). Almost half coal power plants seen unprofitable to operate. *Bloomberg*. Accessible at: https://www.bloomberg.com/news/articles/2018-11-30/almost-half-of-coal-power-plants-seen-unprofitable-to-operate.

Clean Cooking Alliance. (2018). *South Africa*. Accessible at: http://cleancookingalliance.org/country-profiles/37-south-africa.html.

Creamer, T. (2013, May 16). Eskom awards 46 wind-turbine contract for Sere Wind Farm. *Creamer Media's Engineering News*. Accessible at: http://www.engineeringnews.co.za/article/eskom-awards-46-wind-turbine-contract-for-sere-wind-farm-2013-05-16/rep_id:4136.

CSIR. (2015a, January 21). *2014 sees financial benefits of renewable energy exceed costs in South Africa*. Accessible at: https://www.csir.co.za/2014-sees-financial-benefits-renewable-energy-exceed-costs-south-africa.

CSIR. (2015b, August 19). *First half of 2015 sees financial benefits from renewable energy with huge cost savings*. Accessible at: https://www.csir.co.za/first-half-2015-sees-financial-benefits-renewable-energy-huge-cost-savings.

Department of Mineral Resources (DMR). (2014). South Africa's coal industry—Overview, 2014 (Report R111). Accessible at: http://www.dmr.gov.za/LinkClick.aspx?fileticket=uePTS1drX6A%3D&portalid=0.

Department of Water Affairs and Forestry (DWAF). (2009). *Water resources analysis, Vaal River system: Large bulk water supply reconciliation strategy* (DWAF Report No. PRSA C000/00/4406/05). Pretoria: DWAF.

Deutsch, K. (1961). Security communities. In J. Roseau (Ed.), *International politics and foreign policy: A reader in research and theory*. Glencoe, NY: Free Press.

DPRU. (2018, June 2). *Monitoring the performance of the South African labour market: An overview of the South African labour market for the year ending 2017 quarter 2*. Development Policy Research Unit, University of Cape Town. Accessible at: http://www.dpru.uct.ac.za/sites/default/files/image_tool/images/36/Publications/Other/2018-06-13%20Factsheet%2021%20-%20Year%20ended%202017Q2.pdf.

DWA. (2012). *Full technical report on the implication of climate change impacts on water resources planning in South Africa*. Pretoria: Department of Water Affairs.

Edigheji, O. (Ed.). (2010). *Constructing a developmental state in South Africa*. Cape Town: HSRC Press.

EIA. (2016). *Global energy intensity continues to decline*. Washington, DC: US Energy Information Administration. Accessible at: https://www.eia.gov/todayinenergy/detail.php?id=27032.

Eskom. (2005). *Annual Report, 2004–2005.* Accessible at: http://www.eskom.co. za/sites/heritage/Annual%20Reports/2004-2005%20Annual%20Report.pdf.

Evans, P. (2010). Constructing the 21st century developmental state: Potentialities and pitfalls. In O. Edighej (Ed.), *Constructing a democratic developmental state in South Africa: Potentials and challenges* (pp. 37–59). Cape Town: HSRC Press.

Fine, B., & Rustomjee, Z. Z. R. (1996). *The political economy of South Africa: From Minerals Energy Complex to Industrialisation.* Boulder, CO: Westview.

Geels, F. (2018). Disruption and low-carbon system transformation: Progress and new challenges in socio-technical transitions research and the multi-level perspective. *Energy Research & Social Science, 37,* 224–231.

Geels, F. W. (2005). *Technological Transitions and System Innovations: A Co-Evolutionary and Socio-Technical analysis.* Edward Elgar Publishing.

Geels, F. W. (2010). Ontologies, socio-technical transitions (to sustainability), and the multi-level perspective. *Research policy, 39*(4), 495–510.

Geels, F. W. (2014). Regime resistance against low-carbon transitions: Introducing politics and power into the multi-level perspective. *Theory, Culture & Society, 31*(5), 21–40.

GHS-SA. (2016). General household survey (GHS) 2016 report. *StatsSA.* Accessible at: http://www.statssa.gov.za/publications/P0318/P03182017.pdf.

Gibbs, A., & Lloyd, T. (2018, October 25). *The myth of "clean coal": Why coal can only ever be dirty.* EE Publishers. Accessible at: http://www.ee. co.za/article/the-myth-of-clean-coal-why-coal-can-only-ever-be-dirty.html#. W9KzDxi6I0M.

Gleason, D. (2013, July 10). Medupi farce may cost trillion-plus. *Business Day Live.* Accessible at: https://www.businesslive.co.za/bd/opinion/columnists/2013-07-10-medupi-farce-may-cost-trillion-plus/.

Grubler, A. (2012). Energy transitions research insights and cautionary tales. *Energy Policy, 50,* 8–16.

Harrington, R. (2016, October 9). Where Hillary Clinton stands on climate change. *Business Insider.* Accessible at: https://www.businessinsider.com/hillary-clinton-environment-climate-change-platforms-policies-plans-2016-10? r=US&IR=T.

Hartnady, C. (2010). South Africa's diminishing coal reserves. *South African Journal of Science, 106*(9–10). Accessible at: http://www.sajs.co.za.

Heintz, J. (2010). The social structure of accumulation in South Africa. In T. McDonald, M. Reich, & D. Kotz (Eds.), *Contemporary capitalism and its crises* (pp. 267–285). New York: Cambridge University Press.

IEA. (2006). *World energy outlook.* Accessible at: https://www.iea.org/publications/freepublications/publication/weo2006.pdf.

IEA. (2017). *World energy investment 2017.* Accessible at: https://www.iea.org/publications/wei2017/.

InfluenceMap. (2018, December). Who Owns the World's Fossil Fuels? A forensic look at the operators and shareholders of fossil fuel companies. *InfluenceMap*. Accessible at: https://influencemap.org/finance-map.

IPCC. (2007). Climate change 2007: Impacts, adaptation and vulnerability. *Working Group II Contribution to the Fourth Assessment Report of the Intergovernmental Panel on Climate Change*. Accessible at: https://www.ipcc.ch/site/assets/uploads/2018/03/ar4_wg2_full_report.pdf.

IRENA. (2018). *Renewable power generation costs in 2017*. Abu Dhabi: International Renewable Energy Agency. Accessible at: https://www.irena.org/-/media/Files/IRENA/Agency/Publication/2018/Jan/IRENA_2017_Power_Costs_2018.pdf.

Jaglin, S., & Dubresson, A. (2016). *ESKOM: Electricity and Technopolitics in South Africa*. Cape Town: Juta and Company (Pty) Ltd.

Jacobson, M. Z., Delucchi, M. A., Bauer, Z. A., Goodman, S. C., Chapman, W. E., Cameron, M. A., et al. (2017). 100% clean and renewable wind, water, and sunlight all-sector energy roadmaps for 139 countries of the world. *Joule, 1*(1), 108–121.

Kalecki, M. (1971). Class struggle and the distribution of national income. *Kyklos, 24*(1), 1–9.

Kohli, A. (2006). *State and development*. Cheltenham: Edward Elgar.

Lawrence, A. (2013). Neoliberalism, mineral resource governance and developmental states: South Africa in comparative perspective. In F. Bourgouin & J. Nem Singh (Eds.), *The political economy of extraction* (pp. 40–60). New York: Palgrave.

Lawrence, A. (2014). *Employer and worker and collective action: A comparative study of Germany, South Africa, and the United States*. New York: Cambridge University Press.

Lawrence, A. (2018). How can energy and climate justice claims be reconciled? In A. Goldthau, M. Keating, & C. Kuzemko (Eds.), *Handbook on the IPE of energy and resources* (pp. 227–237). Cheltenham: Edward Elgar.

Liefferink, M. (2016, November). *Rehabilitation of mine contaminated wetlands, eco-systems and receptor dams: A just transition to a low carbon economy to combat unemployment and climate change*. Cape Town: AIDC.

Lippit, V. (2010). Social structure of accumulation theory. In T. McDonough, M. Reich, & D. Kotz (Eds.), *Contemporary capitalism and its crises: Social structure of accumulation theory for the twenty first century* (pp. 45–71). New York: Cambridge University Press.

Loorbach, D. (2007). *Transition management: New mode of governance for sustainable development*. Utrecht, The Netherlands: International Books.

Mail & Guardian. (2010, April 3). Zille says ANC stands to make R1bn from Medupi. *Mail & Guardian*. Accessible at: https://mg.co.za/article/2010-04-03-zille-says-anc-stands-to-make-r1bn-from-medupi.

Manuel, T. (2011). Our diagnosis of SA's problems. *PoliticsWeb*. Accessible at: http://www.politicsweb.co.za/documents/our-diagnosis-of-sas-problems–trevor-manuel.

Martinez, D. M., & Ebenhack, B. W. (2008). Understanding the role of energy consumption in human development through the use of saturation phenomena. *Energy Policy, 36*(4), 1430–1435.

McKay, D. (2018, November 29). Exxaro was "frustrated" with Eskom coal procurement, but signs of hope detected. *Miningmx*. Accessible at: https://www.miningmx.com/news/energy/35291-exxaro-was-frustrated-with-eskom-coal-procurement-but-signs-of-hope-detected/.

Mining Review. (2018, February 9). The future of "dirty" coal in South Africa vs reliable renewable energy sources. *Mining Review*. Accessible at: https://www.miningreview.com/future-dirty-coal-south-africa-reliable-renewable-energy-sources/.

Mohamed, S. (2010). The state of the South African economy. In J. Daniel, P. Naidoo, D. Pillay, & R. Southall (Eds.), *New South African review 1: Development or decline?* (pp. 39–65). Johannesburg: Wits University Press.

Montel. (2018). *South Africa Q1 thermal coal exports rise 6%*. Accessible at: https://www.montel.no/fr/story/south-africa-q1-thermal-coal-exports-rise-6/897919.

Monyei, C. G., Sovacool, B. K., Brown, M. A., Jenkins, K. E., Viriri, S., & Li, Y. (2019, January–February). Justice, poverty, and electricity decarbonization. *Electricity Journal, 32*(1), 47–51.

Muller, M. (2013). The regulation of network infrastructure beyond the Washington consensus. *Development Southern Africa, 30*(4–5), 674–686.

Mundy, S. (2019, January 1). India's renewable rush puts coal on the back burner. *Financial Times*. Accessible at: https://www.ft.com/content/b8d24c94-fde7-11e8-aebf-99e208d3e521.

Nathan, L. (2006). Domestic instability and security communities. *European Journal of International Relations, 12*(2), 275–299.

Newell, P. (2018). Trasformismo or transformation? The global political economy of energy transitions. *Review of International Political Economy*, 1–24. https://doi.org/10.1080/09692290.2018.1511448.

NPC. (2012). *National Planning Commission, National Development Plan 2030: Our future—Make it work*. Pretoria: Government Printers.

Ochieng, G., & Nkwonta, O. (2010, November 18). Impacts of mining on water resources in South Africa: A review. *Scientific Research and Essays, 5*(22), 3351–3357.

Pollin, R. (2015). *Greening the global economy*. Cambridge: MIT Press.

Rademeyer, J. (2013). Claim that 94% in SA have access to safe drinking water…Doesn't hold water. *AfricaCheck*. Accessible at: https://africacheck.

org/reports/claim-that-94-of-south-aclaim-that-94-in-sa-have-access-to-safe-drinking-water-doesnt-hold-water/.

Ravallion, M. (2007). Inequality is bad for the poor. In S. Jenkins & J. Micklewright (Eds.), *Inequality and poverty re-examined*. Oxford: Oxford University Press.

SADHS. (2017, May). *South Africa Demographic and Health Survey 2016 key indicators report 03-00-09*. Pretoria: Statistics South Africa. Accessible at: https://dhsprogram.com/pubs/pdf/PR84/PR84.pdf.

SAMI. (2017, October 24). *South Africa's coal exports: Where is it going?* Blog. Accessible at: https://www.southafricanmi.com/blog-24oct2017.html.

SAMI. (2018). *GDP growth calculator*. South African Market Insights. Accessible at: https://www.southafricanmi.com/gdp-growth-estimator.html.

Scarse, I., & Smith, A. (2009). The non-politics of managing low carbon sociotechnical transitions. *Environmental Politics, 18*(5), 707–726.

Scholvin, S. (2014). South Africa's energy policy: Constrained by nature and path dependency. *Journal of Southern African Studies, 40*(1), 185–202. https://doi.org/10.1080/03057070.2014.889361.

Shove, E., & Walker, G. (2007). CAUTION! Transitions ahead: Politics, practice, and sustainable transition management. *Environment and Planning a, 39*(4), 763–770.

Smil, V. (2006). *Energy*. Oxford: OneWorld.

Smith, A., Stirling, A., & Berkhout, F. (2005). The governance of sustainable sociotechnical transitions. *Research Policy, 34*(10), 1491–1510.

SourceWatch. (2018a). *Medupi Power Station*. Accessible at: https://www.sourcewatch.org/index.php/Medupi_Power_Station.

SourceWatch. (2018b). *Kusile Power Station*. Accessible at: https://www.sourcewatch.org/index.php/Kusile_Power_Station.

Starr, C. (1999). Observations on the future of nuclear power and how to get there. In B. Kursunoglu, S. Mintz, & A. Perlmutter (Eds.), *Preparing the ground for renewal of nuclear power* (pp. 29–34). Boston, MA: Springer.

StatsSA. (2017). *Non-financial census of municipalities report, 2017*. Accessible at: http://www.statssa.gov.za/?p=11199.

Steenkamp, L., Lategan, R., & Raubenheimer, J. (2016). Moderate malnutrition in children aged five years and younger in South Africa: Are wasting or stunting being treated? *South African Journal of Clinical Nutrition, 29*(1), 27–31.

Stockhammer, E. (2015). Rising inequality as a cause of the present crisis. *Cambridge Journal of Economics, 39*(3), 935–958.

Swilling, M., & Annecke, E. (Eds.). (2012). *Just transitions: Explorations of sustainability in an unfair world*. Cape Town: UCT Press.

Swilling, M., Musango, J., & Wakeford, J. (2015). Developmental states and sustainability transitions: Prospects of a just transition in South Africa. *Journal of Environmental Policy & Planning*. https://doi.org/10.1080/1523908x.2015.1107716.

Trading Economics. (2018). *South Africa: GDP growth rate.* Accessible at: https://tradingeconomics.com/south-africa/gdp-growth-annual.

Turton, A. (2009). South African water and mining policy: A study of strategies for transition management. In D. Huitema & S. Meijerink (Eds.), *Water policy entrepreneurs: A research companion to water transitions around the globe* (pp. 195–214). Cheltenham, UK: Edward Elgar.

UBS Investment Research. (2013, January 2). *Entergy Corp.: Re-assessing cash flows from the Nukes.*

UNDESA. (2014). *International decade for action 'Water for Life' 2005–2015.* Accessible at: http://www.un.org/waterforlifedecade/scarcity.shtml.

UN Water. (2006). *Water a Shared Responsibility.* Accessible at: http://unesdoc.unesco.org/images/0014/001454/145405e.pdf#page=519.

Van Wyk, J. J., Rademeyer, J. I., & Van Rooyen, J. A. (2010, November). *Position statement on the Vaal River system and acid mine drainage.* Department of Water Affairs. Accessible at: http://www.dwaf.gov.za/Projects/AMDFSLTS/Documents/Vaal%20River%20System%20&%20AMD%20Version%203.pdf.

Vardi, I. (2016, March 7). Hillary Clinton Showed Support, Associates Profited from Ex-Im Bank Financing World's Largest Coal Plants in South Africa. *Desmog.* Accessible at: https://www.desmogblog.com/2016/03/07/hillary-clinton-showed-support-associates-profited-building-world-s-largest-coal-plantssouth-africa.

Wakeford, J. (2012). *Socioeconomic implications of global oil depletion for South Africa: Vulnerabilities, impacts and transition to sustainability* (PhD thesis). Stellenbosch University, Stellenbosch.

Ward, C. (2018, March 14). Biofuel from sugarcane—Why is SA not rushing ahead? *Daily Maverick.* Accessible at: https://www.dailymaverick.co.za/article/2018-03-14-op-ed-biofuel-from-sugarcane-why-is-sa-not-rushing-ahead/.

Wassung, N. (2010). *Water scarcity and electricity generation in South Africa* (MPhil thesis). School of Public Management and Planning, University of Stellenbosch. Accessible at: http://hdl.handle.net/10019.1/18158.

Webster, E. (2017, December 11). South Africa needs a fresh approach to its stubbornly high levels of inequality. *The Conversation.* Accessible at: https://theconversation.com/south-africa-needs-a-fresh-approach-to-its-stubbornly-high-levels-of-inequality-87215.

Weisskopf, T. E. (1979, December). Marxian CRISIS theory and the rate of profit in the postwar U.S. economy. *Cambridge Journal of Economics, 3*(4), 341–378.

WikiLeaks. (2010). *Eskom and the World Bank loan for Medupi.* Accessible at: https://wikileaks.org/plusd/cables/10PRETORIA125_a.html.

Wilson, E., Stephens, J., & Peterson, T. (2015). *Smart grid (R)evolution: Electric power struggles.* New York: Cambridge University Press.

World Bank. (2019). *Access to electricity—South Africa*. Washington, DC: World Bank Group. Accessible at: https://data.worldbank.org/indicator/EG.ELC.ACCS.ZS?locations=ZA.

Yelland, C. (2015, January 24). Eskom's Boiler contractor at Medupi and Kusile Defends its reputation. *Moneyweb*. Accessible at: https://www.moneyweb.co.za/uncategorized/eskoms-boiler-contractor-at-medupi-and-kusile-defe-2/.

CHAPTER 3

Eskom and the Dual Character of the South African State

Abstract South African policymakers' failure to embrace a rapid transition to renewable energy (RE) generation results not simply from insufficient appreciation of its immediate and longer term benefits, but from their protracted deterrence of this option. The sources and mechanisms of deterrence, however, long predate the advent of RE competition, as may be seen from the evolution of South Africa's energy parastatal, Eskom. Prior to the current era of universal franchise, the combination of lengthy periods of dominant party rule with the rise of dominant mining interests made parastatals a useful means of combining industrialisation with political patronage. At the same time, however, Eskom's governance structures were explicitly designed to insulate it from popular pressure and demands for transparency in all but the most extraordinary of circumstances—a tendency that continued into the democratic era. The use of "preferential procurement" legislation to govern licencing and government purchasing creates incentives for companies to recruit the politically well-connected onto their boards, increasing the interpenetration of public and private interests. These factors combined to make the neo-patrimonial governance style of the Zuma presidency more likely, and conflicts of interest virtually inevitable. In parallel with these developments, however, an absence of transparency and accountability—and the growing challenges of sourcing affordable coal for electricity generation—have conspired to make inadequate infrastructure planning and resultant power outages also increasingly likely. The "perfect storm" of "peak coal," patronage, poor planning and lack of accountability

© The Author(s) 2020   59
A. Lawrence, *South Africa's Energy Transition*, Progressive Energy
Policy, https://doi.org/10.1007/978-3-030-18903-7_3

have combined to plunge Eskom into a protracted crisis from which it is unlikely to emerge intact.

Keywords   Eskom · Corruption · State theory · State capture · Blackout · Load-shedding

South African policymakers' failure to embrace a rapid transition to renewable energy (RE) generation results not simply from insufficient appreciation of its immediate and longer term benefits, but from their protracted deterrence of this option. This posture can be seen as a hidden strand of continuity beneath the seeming turbulence, even chaos, surrounding Eskom's governance over the past decade. Understanding the sources of both the continuity as well as the turbulence of South Africa's energy policy making in turn requires a brief elucidation of the institutional structures and political cultures of the South African state. Resulting from waves of popular contestation, repression, and accommodation, these structures also reveal a dual character. Whereas the establishment of universal franchise is a relatively recent development, the pattern of lengthy stretches of dominant party rule—interspersed with brief episodes of coalition or pact governments, and characterized by powerful, often charismatic, heads of state presiding over quiescent parliaments and networks of patronage—is a much older norm.[1] The constitutional, rule of law state is a palimpsest in which traces of the older, less formalized political culture are discernible.

Although by no means unique—or even atypical—this dual character of South Africa's state is less apparent during periods of growth or regime stabilization, and more evident during periods of political or economic (or, as at present, political and economic) turbulence or crisis. Among the notable characteristics of the Zuma presidency, for example, has been the rise of neo-patrimonial governance (manifest in increased factionalism and a neo-traditional, patriarchal, identity-based cultural politics). With reference to the shifting norms within the ruling African National Congress (ANC), Lodge has argued that this governance style has deep historical roots and several enabling factors.[2] The colonial and apartheid states criminalized large swathes of everyday African life, thus blurring the line between legal

---

[1] See, e.g., Van Vuuren (2006) and Ellis (2013).

[2] Lodge (2014).

and illegal. With most of the population forced to cross this line at various points in their lives, ANC activists found able recruits among township gang members, such as Joe Modise, who started as a car thief and smuggler in his youth, went on to lead Umkhonto we Sizwe (the armed wing of the ANC) in exile, and later became Mandela's Minister of Defense and beneficiary of the multi-billion rand 1997 arms contract. Zuma, while head of the ANC's intelligence department, similarly enlisted Johannesburg-based criminal syndicates to procure arms. Mass (especially youth) unemployment sustains an aspirational culture of conspicuous consumption and ensures the continued power of the ANC Youth League within the party; it also accentuates perceptions of statutory welfare as forms of patronage.

To be sure, this tendency has always conflicted—and still contends—with more radical currents, such as those that see the Freedom Charter's insistence upon equitable sharing of the country's wealth and benefits as fundamental. Arguably the biggest factor tilting norms in a parasitic direction, however, has been the role played by the Minerals Energy Complex (MEC) political economy, in which licencing and government purchasing is regulated by "preferential procurement" legislation, creating incentives for companies to recruit the politically well-connected onto their boards.[3] At the heart of this dynamic lies the role of Eskom in providing a steady supply of political patronage at least as reliable as its ostensible role of providing a steady supply of electricity. Despite major changes over the course of its nearly century-old existence, it remains the largest state-owned enterprise (SOE), and arguably, its economic and societal importance is greater than that of any other SOE, and its functions more varied. Its size is noteworthy in global terms as well—not only is it Africa's largest electricity producer (providing 38% of all electricity consumed on the continent) and the world's eleventh largest in terms of installed capacity, but it also was recently ranked the sixth largest African company.[4] Ostensibly, at least, it has also managed to buck the fate of full privatization that has befallen most other energy parastatals in the OECD and in much of the Global South—although as discussed below, its shareholder governance reforms have moved it closer to this end.[5]

---

[3] Lodge (2014).

[4] Jaglin and Dubresson (2016: 1).

[5] Eberhard (2005).

Yet the crisis over Eskom's governance has become so pronounced in recent years that it is an open question whether the public trust it formerly enjoyed will ever be regained, thereby allowing its political and economic centrality (if not also its public status) to endure. One clear symptom of political manipulation has been its executive turnover rate: two years before Zuma's presidency had ended, it witnessed four board chairs, six CEOs and five chief financial officers at Eskom; similarly, four ministers took office at the Ministry of Energy, and five at the Department of Public Enterprises.[6] This crisis is usually characterized as part of a broader pattern of "state capture" or corruption on a grand scale. Too often overlooked, however, is the question of location and quality of corruption, pointing to the distinction between bureaucratic corruption (concerning the everyday enforcement of policies and laws) and political corruption (concerning the formation of policies and laws, in which policymakers exercise much broader latitude).[7] South Africa has enacted several measures against the former, such as the 1999 Public Finance Management Act, requiring disclosure of "any direct or indirect personal or private business interest that [a given] member or any spouse, partner or close family member may have in any matter before the accounting authority."[8] As this chapter details, Eskom's recent trajectory straddles both types.

Discussions of Eskom and corruption have tended to focus on the "Zuptas"—the portmanteau term combining Zuma (and members of his family) with his close associates, the Gupta brothers—as well as several others associated with them. Over the past decade, the network used tens of billions of rands worth of contracts for transport parastatal Transnet's locomotives purchases, and Eskom-related coal and consulting contracts, to funnel billions of rands into shell corporations or inflated fees.[9] Crucial transnational assistance was provided by major firms such as UK PR firm Bell Pottinger (whose prior clients include Augusto Pinochet, the CIA, and Syria's ruling Assad dynasty, and whose South African activities led to its collapse in 2017) and the world's largest consultancy firm McKinsey & Co.[10]

---

[6]Styan (2015: 54).

[7]Bratsis (2014: 110).

[8]PFMA (1999: Section 50[2] and [3]).

[9]Basson and Du Toit (2017), Hofstatter (2018), and Pauw (2017).

[10]See, e.g., Basson and Du Toit (2017: 231–241, 270–272) and Hofstatter (2018: 61–67, 230–232).

A common implication is that a few bad apples allowed their greed to subvert public service in the public good; the remedy is therefore to replace bad apples with good apples and continue as before.[11] This view presumes that, being both clearly reprehensible and few in number, wrong-doers may be straightforwardly identified and replaced. It is thus blind to more systemic factors and their broader enabling culture. Yet as Justice Malala argued with reference to (then Deputy President, now President) Cyril Ramaphosa, "he has had to defend every unprincipled decision taken by [Zuma]" and "preside over the mess and reports of wanton corruption at Eskom and allegations of impropriety at almost every level." Should not "those who see in him a Messiah of sorts begin to ask themselves what he can do as Number 1 that he could not do as Number 2? … The saddest part of looking to Ramaphosa as a potentially redemptive leader is that we have also lost agency as a people."[12]

Less often, discussions of corruption refer back to the R30 billion arms deal in which not only Zuma, but also former presidents Mbeki and Man-dela, among others, may be implicated.[13] An older, comparative perspec-tive would see these cases as typical of "modernizing" societies.[14] Still less explored is a broader global perspective, one that sees the South African case as "just one quite typical example of a global trend in the growth of increasingly authoritarian, neopatrimonial regimes where a symbiotic rela-tionship between the constitutional and shadow states is maintained, but with real power shifting increasingly into the networks that comprise the shadow state."[15]

Rarer still, perhaps, are critiques underscoring the interconnections among crises associated with economic, ecological, and political unsus-tainability. Sterling political leadership of the highest probity that does not diversify the economy away from mining will, *ceteris paribus*, still

---

[11] For example, Hofstatter concludes by stating "The plunder of Eskom and other paras-tatals brought the economy to the brink of collapse. … Without Gigaba, Molefe, Singh, Brown, Pamensky, Koko, and Maritz, the looting game could never have been played to such devastating effect. That they are gone is small solace for taxpayers forced to foot the R50-billion bill. But for now, at least, the licence to loot has been revoked" (Hofstatter 2018: 244).

[12] Malala (2015: 147–148).

[13] Feinstein (2007).

[14] Huntington (1968).

[15] Bhorat and Swilling (2017).

yield extreme inequality and accelerated capital flight, and thus the near-inevitability of increased political instability. A continued reliance on coal generation inevitably yields increased ecological degradation, itself contributing to the near-inevitability of increased political instability.[16]

Analogously, it is also important to situate Eskom in its broader institutional context. With specific reference to the energy sector, the Department of Energy (DoE), established in 2009 (when the Department of Minerals and Energy was divided into the DoE and Department of Mineral Resources) is required by law to produce an Integrated Resource Plan (IRP) governing the generation of electricity—although in practice the drafting of the IRP is regularly outsourced to Eskom planners. No new capacity can be licenced that is not explicitly referenced in the current IRP. Eskom reports to the Department of Public Enterprises and pays dividends to the Treasury, but its prices are set by the National Energy Regulator (Nersa), established in 2004, replacing the 1995 body, the National Electricity Regulator (NER), with an additional remit to regulate nuclear as well as fossil fuel energy. In addition to generating approximately 95% of the country's electricity supply, Eskom as exclusive large-scale vendor also exercises a monopsony of power. It also owns the transmission grid, and is responsible for 60% of distribution; municipalities sell the balance on to local consumers. It is thus a quintessential exemplar of large-scale centralization of energy purchasing, generation, supply, and transmission.[17] The centralization of diverse essential services—not subject to continuous public scrutiny or market pressure—arguably makes corruption no less than suboptimal governance more likely. Normatively, the question of systemic reform (addressed in the concluding chapter) becomes therefore how to combine increased transparency, accountability, and security of supply with increased energy justice and access for the poorest South Africans.

In describing its evolution from the interwar period to the postapartheid era, this chapter details how Eskom's bureaucratic autonomy arose as the state's response to working class contention while catering to dominant mining interests. This autonomy, however, became a perpetual source of weakness in Eskom's capacity to continuously plan and adjust for energy demand. The advent of apartheid meant that Eskom, together with the other SOEs, became an irresistible vehicle for the ruling National Party

[16]But see Satgar (2018).
[17]Eberhard (2007).

(NP) to exercise patronage in staffing and mining contracts; any resulting inefficiencies were glossed over by the highly favourable external environment of the "golden quarter century" of global growth and strong foreign direct investment, above all, from the United States. Less successfully, the NP used Eskom as well as a means for pursuing energy autonomy and inward import-substituted industrialisation. With the advent of democratization, the ANC largely echoed this pattern. It has used SOEs as tools of patronage, momentarily buttressed by the minerals boom induced by Chinese demand from the late 1990s to the global financial crisis. Jettisoning the NP's tariffs and currency controls, it has also failed to steer the economy onto a new growth path, indeed, weakening the country's manufacturing base, increasing fiscal instability and exposure to currency markets, and reducing non-mining sectors' international competitiveness. As a popular referent of current malaise, "state capture" has served to underscore interrelations among economic, ecological, and political fragilities.

From MEC Origins to Apartheid Crises

South Africa was once a trailblazer in electrification: in 1881, barely two years after Edison invented the incandescent lamp, South Africa's first electric lights were installed at the Cape Town railway station.[18] The following year, the booming diamond mining town Kimberley became the first city in Africa and the southern hemisphere to use electric street lights, shortly followed by Cape Town, ahead of London. Yet the legacy of this "advantage of backwardness" would not endure. In the wake of the Union of South Africa's establishment in 1910, with a population of about 6,000,000, the country's generating capacity totalled fewer than 200 MW, reaching only a small minority of businesses and the general population.[19]

The 1919 syndicalist strike wave that included momentary control by striking workers of the city of Johannesburg led the South African Railways general manager to vow never to buy power from municipalities because of the spectre of their control by "bolshevism"—a fear only strengthened by the 1922 Rand miners' strike. The solution therefore was to establish a parastatal independent of not only local, but also largely of parliamentary and governmental control. The 1922 Electricity Act therefore established

---

[18] Eberhard (2007: 216).
[19] Christie (1984: 5–6).

the Electricity Control Board (ECB) that would control and licence supply and set prices; and the tax-exempt Electricity Supply Commission (Escom), inaugurated in 1923, for the provision and distribution of cheap and abundant electricity. Following from the gold standard's abandonment, liberating the price of gold in the 1930s and expansion during World War II, Escom's generating capacity had grown five-fold.[20] Escom's mandate was to provide electricity as cheaply, continuously and securely as possible, passing its savings on to its customers—initially, the Victoria Falls and Transvaal Power Company (VFTPC), a privately held foreign company that sold electricity to the gold mines at an immense profit before being bought out by Escom in 1948.

Escom funded its growth through loans and set electricity prices based on loan repayments and anticipated future sales—a financing mechanism prone to under- or over-investment miscalculation. This weakness was apparent when the rapid expansion of gold mining in the late 1940s led to several power outages and load shedding from 1948–1953. At the same time, Afrikaner coal mine owners—now an important constituency of the ruling NP—chafed under restrictions against direct export of their (mostly lower grade) coal that their Anglophone competitor mine owners did not face. Escom responded over the next two decades with a program of construction of the Komati, Camden, Grootvlei, Hendrina, and Arnot power stations (with a cumulative generating capacity of nearly 7500 MW), and equipment for several others, funded with tens of millions of dollars in World Bank loans. In the three decades after WW II, capacity increased six-fold, to nearly 15,000 MW. Yet the reserve margin rarely exceeded the modest level of 10% throughout this period, even dipping close to 5% in the mid-1960s.[21] Subsequent fiscal constraints led it to establish a capital development fund that allowed Escom to retain earnings and use profits for construction projects, independent of government control or approval. This power enabled Escom to respond to the oil crisis by almost doubling prices from 12 cents/kWh in 1970 to more than 20 cents/kWh in 1975, which in turn allowed it to initiate construction of Africa's first and only nuclear power plant (and related secret nuclear weapons program) at Koeberg.[22]

[20] Styan (2015: 10–12).

[21] Eberhard (2007: 219).

[22] Eberhard (2007: 222); Styan (2015: 13–14).

Several factors induced Escom's managers to opt for maximal capacity expansion: forced to contend with ever-lower grades of coal while encountering difficulties in integrating larger 600 MW plants into the grid, they also sought to assuage apartheid ideologues' paranoid fears of "total onslaught" in the wake of the 1976 Soweto uprising, and wars with the liberation movements in Angola and Mozambique. In 1982, they forecast a R62 billion construction program aiming to increase capacity to 70,000 MW by 2000, and the next year had further capacity of 22,260 MW on order or under construction, financed by year-on-year tariff increases above the rate of inflation and increased reliance on both domestic and international loans, which in 1982 alone totalled nearly R2 billion.[23] In all, between 1970 and 1994, Eskom added 33,910 MW of new generation capacity. The fact that actual peak demand in 2000 did not reach even 30,000 MW showed how far off kilter the parastatal's planning had become.[24]

Political fallout from the rising costs led to an official investigation by the De Villiers Commission from 1983–1985, which instigated a change of management and management structure (now divided into a management board and an electricity council in charge of policy, with representatives from industry as members), a change of name to Eskom (from its Afrikaans name *Elektrisiteitsvoorsieningskommissie*) in 1987, a jettisoning of the parastatal's not-for-profit principle (despite retaining exemption from taxes and dividends), and a moratorium on new plant construction that would endure over the next two decades.[25] This decision laid the groundwork for Eskom's policy of price decreases—20% between 1992 and 1996, and a further 15% reduction between 1994 and 2000—that fostered the establishment of energy-intensive metal industries such as mining multinational BHP Billiton's Hillside, Bayside, and Mozal plants (based in Mozambique but dependent on Eskom-generated electricity).[26]

One legacy of these reductions—unearthed in 2009 by investigative journalist Jan de Lange after remaining hidden from public knowledge for years—is the lower rate that Eskom charges its biggest (mostly mining and industry) customers compared with what most other consumers pay. In

[23]Padayachee (1991: 96).

[24]Styan (2015: 15).

[25]Styan (2015: 16).

[26]Eberhard (2007: 234).

2011, these firms accounted for about a third of Eskom's total revenues, while consuming over 40% of its total electricity produced. They were bene-ficiaries of generous 30-year contracts, and linked (in the case of aluminium smelters) to fluctuations in the price of aluminium. Their official rationale of jobs creation begged several questions. As Chris Yelland asked at the time,

> Why are the details of the deals kept secret, while all other domestic, commer-cial, agricultural, industrial and mining customers pay transparent tariffs that are openly published? Why should a few foreign companies get electricity at below cost, while local customers face massive increases that effectively sub-sidise the losses Eskom incurs on the secret deals? Why should thousands of GWh of locally produced electricity be sold below cost for export by a foreign-owned company in the form of aluminium ingots, while security of supply in South Africa is threatened and local industry is starved of electricity? Does it really add value to the South African economy when bauxite is mined and refined to alumina elsewhere, then shipped to South Africa to take advantage of subsidised electricity purchased below cost to convert it into aluminium ingots for export? Does aluminium production in this way really contribute to jobs in South Africa, when staffing at the smelters is relatively low and there are no upstream and few downstream value-adding activities?[27]

In other words, like the VAT, the state-controlled electricity market amounts to a substantial regressive tax. In 2013, when the average gen-eration cost to Eskom was about 40c/kWh and industry customers paid close to this amount, BHP Billiton, which buys 9% of Eskom's electricity, paid only 22.65c/kWh, while 4.5 million residential customers paid on average R1.40/kWh—more than six times more.[28] Moreover, such deals created a precedent and thus could become the rationale for preferential deals of any kind—a pattern that became especially manifest during the Zuma presidency.

Concerns about governance competency and accountability at Eskom and other SOEs persisted after democratization in 1994, motivating even more potentially far-reaching reforms. The 1998 White Paper on Energy Policy recommended ending Eskom's (near) monopolies on gener-ation, distribution, and transmission,[29] while a subsequent cabinet memo

[27]Yelland (2011).

[28]ENCA (2013).

[29]DME (1998).

announced that 30% of electricity supply would be generated by independent power producers (IPPs). Then-Minister of Public Enterprises Jeff Radebe alluded to possible motivating factors in 2000:

> in many instances, [SOEs'] traditional methods of operation and business management, human resource development, and even their targeted constituency base were inappropriate in a democracy. Furthermore, they command large procurement budgets, social and human resources and some of them dominate economic sectors in which they operate. ... In this context, the accelerated restructuring of state-owned enterprises to meet these combined challenges has become a matter of urgency.[30]

He reiterated the White Paper's findings that commercial energy was unlikely to become scarce in the short or medium term, that energy security could be strengthened through increased diversification and flexibility of supply, and that national goals ensuring access can be achieved through regulating delivery rather than direct provision. He further observed that "the reduction of emissions is becoming increasingly important," and global energy trends included "energy efficiency, demand-side management, ... and [the promotion of] research and development of alternative and renewable energy sources." For these reasons, the government would oversee the creation of new regional electricity distributors (REDs) to achieve better service and lower prices; the unbundling of Eskom's transmission, distribution, and generation functions into separate corporate entities to improve electricity supply's transparency and accountability; and to promote internal generating competition, prior to the introduction of private sector competition in generation, to improve efficiency and ensure the continued rollout of electricity to poor urban and rural communities.[31]

The Cabinet approved the broad thrust of these reforms in May 2001 under the rubric of "managed liberalization." This included Eskom's vertical unbundling, with the establishment of a separate state-owned Transmission Company; a multi-market model electricity market framework; and the diversification of primary energy sources with the inclusion of IPPs. The reforms were followed in November 2003 with the White Paper on RE, stipulating an RE generation target of 10,000 GWh by 2013.[32] Despite

---

[30]MPE (2000: 3).

[31]MPE (2000: 129–134).

[32]DME (2003).

these bold intentions, most of the reforms were never enacted, including establishment of the REDs and unbundling of functions. Indeed, the temptation to retain the power of patronage proved too strong to resist. Instead of the initial proposals to reduce Eskom's market share of generation to 35% as a strong signal to the private sector to invest on equal terms, Eskom's top leadership successfully lobbied to retain 70% of the generation market "in order to meet Government's developmental objectives".

As then-Board Member of the National Electricity Regulator (Nersa) Anton Eberhard observed, this decision "is not well understood. ... If the reference to 'development' means electrification, then it does not make sense as Eskom will no longer be [directly] involved .... If it refers to affirmative procurement practices, these conditions could be included in any future privatization deal. ... Investors argue that there is no logic to this policy and that Eskom's share could and should be reduced to below 35 percent."[33] This decision effectively deterred potential investors from taking the risk of engaging in a sector in which the state would be able to control energy prices. Predictably, subsequent attempts by Eskom to promote IPPs, such as the Pilot National Cogeneration Programme, the Medium-Term Power Purchase Programme and the Multisite Base-load Independent Power Producer Programme, all failed.[34]

The 2011 Independent Systems and Market Operator Bill, which reiterated essentially similar reform goals as the 1998 White Paper, met the same fate and was never enacted. Instead, some of the reforms recommended by the De Villiers Commission were reversed under the Eskom Conversion Act of 2002, which transformed Eskom into a public company and reverted it to a single board structure. More importantly, the Act removed Eskom's tax exemption, so that it now paid dividends to the state, represented by the Minister of Public Enterprises, who was designated as the utility's sole shareholder. This meant that Eskom was no longer oriented toward providing only low-cost power to industry and consumers, but instead toward serving as an income generator for the state and municipalities. Dividends rose as a result from R549 million in 2003 to R1.6 billion in 2005/2006.[35] Mismanaged overinvestment in capacity had thereby given rise to mismanaged underinvestment.

---

[33]Eberhard (2007: 250).

[34]DoE (2009) and Yelland (2009).

[35]Styan (2015: 19–20).

POWER OUTRAGES, 2008 AND AFTER: FROM CRISIS OF DELIVERY TO CRISIS OF GOVERNANCE

In May 2006, responding to questions about major blackouts in the Western Cape province, entailing estimated losses of R5.5 billion to area businesses, then President Mbeki attempted to send a reassuring message. "We shouldn't frighten ourselves too much," he declared. "There were regrettable losses suffered by many businesses, but there is no crisis.... Whatever needs to be done to make sure that the economy grows and new investors come into the economy is being done on the energy and other sides." He added, moreover, that the parliamentary opposition's proposal to a set up a commission to probe the government's "failure to meet South Africa's national electricity capacity needs" would serve no useful purpose. When asked by one of his party's MPs when the opposition would ever see the light, he quipped, "they will no doubt plead that they can't see the light when Eskom has switched off the light."[36]

They didn't have long to wait. In the third week of January 2008, the country's biggest blackout struck, as it lost over 20% of its generating capacity—a level which rose to 25% the following week, causing Parliament to declare a national state of emergency on January 25. In the run-up to the blackout, Eskom had already taken 3700 MW of capacity off the system for planned maintenance; unplanned breakdowns (due largely to overreliance on poor-quality coal) had caused an additional loss of 5000 MW.[37] The previous year, an Eskom-commissioned report had warned that the coal supply division was riddled with inefficiencies and incompetence, and power station stockpiles had run down to dangerously low levels. Eskom chief executive Jacob Maroga responded by shelving the report and firing its author.[38]

The blackout's immediate result was massive traffic gridlock, as traffic lights went out; immense levels of food loss, when refrigeration failed; a spike in hospital deaths, including during operations; and a dawning awareness among millions of South Africans about the centrality of electricity access to their everyday lives and well-being. Concomitantly, it laid bare the absence of any alternative for most of them to Eskom's near-monopoly

[36] Le Roux (2006).

[37] CDE (2008: 3).

[38] Hofstatter (2018: 76).

on generation and transmission. Over the coming months, as scheduled blackouts—or, to use Eskom's preferred euphemism, "load-shedding"—became a regular occurrence, once a week or more, lasting several hours each, the longer-term economic toll became evident as thousands of workers were laid off.[39]

The impact of the 2008 crisis has been immense, and may come to be seen as the beginning of the end of Eskom's current *modus operandi*. In the immediate term, the mining sector suffered losses of R600 million per day in export revenues, and the country as a whole, up to R2 billion per day.[40] The longer term cost of the outages was estimated at 10% of GDP by 2014, or roughly R300 billion.[41] Although the commodities boom—led by Chinese demand from 1999 to 2006 (when post-apartheid growth peaked at just over 7%)—had already begun to falter, and the full effects of the global economic crisis inaugurated by Lehman Brothers' collapse in late 2008 had yet to occur (leading to a contraction of −2.60% in the second quarter of 2009), the Eskom crisis marked a political no less than economic turning point.

That this was a chronicle of a death foretold soon became widely apparent, and critics identified a wide array of culprits with varying degrees of plausibility. Michael Spicer, at that time CEO of Business Leadership of South Africa (an employers' association representing South Africa's largest corporate interests), pointed to Eskom's adoption of Black Economic Empowerment (BEE) policy as a major factor. Notwithstanding BEE's lack of success in reducing overall black unemployment and inequality levels, since the policy had existed for less than a decade prior to the 2008 blackout, it is implausible to argue that on its own it was a significant factor.[42] Others blamed poor coal quality or lack of funds. In fact, the government's 1998 White Paper on Energy Policy had predicted a national electricity shortage in 2007, arguing that "the next decision on supply-side investments will probably have to be taken by the end of 1999 to ensure that the electricity needs of the next decade are met."[43] Although from 2001

---

[39] CDE (2008: 1).

[40] Donnelly (2009).

[41] Styan (2015: 1).

[42] He referred to the fact that at the time, Eskom was buying "20 percent of its coal from hundreds of small BEE contractors" that was "expensive" and "often of poor quality" (CDE 2008: 9).

[43] DME (1998: 53).

to 2004 the government had substantially designed blueprints for a competitive electricity market that would encourage private sector investment, as we have seen, it then shelved these plans. Capacity underinvestment was not due to a lack of funds, since from 2000 to 2007, Eskom earned profits on electricity sales each year. Yet by the time Eskom lifted its moratorium on new coal plant investment in 2004, it was too late to see the approved increase in capacity come online.

The need had been apparent many years earlier, for example, in Eskom's declining reserve margin for generation—the difference between maximum generation capacity and current electricity demand. From the Soweto uprising to the first democratic elections in 1994, the reserve margin more than tripled from under 10 to 31%.[44] While the minimum accepted margin—one that would allow for scheduled maintenance and unscheduled breakdowns (such as those occurring in 2007)—is 20%, the actual reserve margin subsequently fell to 15% in 2001, and then to 7% in 2007. At the same time, normal operating capacity fell from its historic level of 90% to 75% (before breakdowns) in 2007.[45] Making matters worse, private sector consumption (especially by large mining and metals companies, the most energy-intensive users) failed to adequately increase efficiencies of production, in large part because the real price of electricity fell from the late 1980s until 2007. Few critics at the time, however, identified overreliance on coal as a more fundamental source of energy insecurity and unreliability—although the grounds for doing so were already well-known by this time.

The 2008 Blackout also served to underscore the backlash against Mbeki's national and party leadership that had been growing since the 1990s. Mbeki had overseen the jettisoning of the Reconstruction and Development Programme (RDP)—which envisioned Keynesian deficit spending to address apartheid's legacies—for his Growth, Employment, and Redistribution (GEAR) programme of fiscal austerity, reduced tariffs and currency liberalization. GEAR can arguably be read as marking continuity from the Botha to Zuma presidencies, albeit one largely induced by a secret $850 million IMF loan approved by President De Klerk with the tacit acceptance of the newly-unbanned ANC.[46] Mbeki had then succeeded in marginalizing his allies and the left wing of his party more broadly; when he dismissed

---

[44] Eberhard (2007: 219).

[45] CDE (2008: 4–7).

[46] Marais (2011: 137) and Kasrils (2018).

his then-deputy president Zuma on allegations of corruption in 2005, his control over the ANC and state government seemed complete.

Yet his deeply mistrustful attitude toward all challengers and his unwillingness to consult and bargain openly contributed to his political downfall at the ANC's 2007 Polokwane conference. Certainly, his posture of AIDS denialism was an additional major factor, cutting short the lives of hundreds of thousands of HIV-positive South Africans, who could have been saved had the universal ARV access policy ultimately adopted been embraced at the start of his presidency. Mbeki argued that poverty, not the HIV virus, causes AIDS; notwithstanding his repeated refusal to engage with medical consensus on this point, he never had a convincing explanation for what caused poverty. This underlies what Habib has convincingly argued was the "central contradiction" of Mbeki's presidency: namely, that in identifying South Africa's two economies (the modern, efficient, internationally competitive metropolitan economy vs. the informal, marginalised, poor, mostly rural economy), he failed to understand that "the policies and functioning of the first 'economy' were precisely what was creating the poverty and immiseration of the second." This failure "enabled him to assume that the first economy required no intervention and that only the second was in need of policy reform and social assistance."[47]

In fact, years before the ANC's unbanning, anti-apartheid scholarship had articulated a strong critique of the "dual economies" perspective. Arguing against the mainstream development framework of Arthur Lewis, Gustav Ranis and John C. H. Fei (also consonant with Mbeki's perspective), Archie Mafeje observed that while apartheid ideology was predicated on denying it, the fundamental basis of colonial and apartheid practice linked rural and African underdevelopment to white, urban industrial development:

> the urban-industrial areas, far from supplementing rural incomes through remittances, are holding the rural areas to ransom. The South African employers for years have refused to pay to their black industrial labour *subsistence wages* [and] ... have obliged the already poverty-stricken rural areas to contribute *one-third of their industrial wage fund*. ... contrary to popular opinion, in South Africa it is the countryside, through labour migration and

---

[47] Habib (2013: 91).

measurable financial exploitation, that supports the cities and not the other way round.[48]

Nonetheless, this insight went unheeded. Although Mbeki's position was more conservative and pro-business than most other voices in the ANC at the time, it was by no means unique; its prevalence was strengthened by the party's dominant tendency, even before the 1994 elections, to demobilize its base and focus on elite-level bargaining and policy formulation.

In any event, the political backlash against Mbeki did not herald a fundamental departure in energy and development policy, and the left wing of the ANC—including the Youth League, South African Communist Party (SACP) and much of Congress of South African Trade Unions (Cosatu)— that had pinned its hopes on Zuma's leadership challenge without independently organizing a credible alternative, would soon experience bitter disappointment. Although the major policy documents of Zuma's early presidency—the *Industrial Policy Action Plan* and *New Growth Plan*— envisaged a more activist state policy than the Mbeki presidency had articulated, neither adequately addressed the entwined problems of growing inequality, unemployment, and environmental degradation. Despite the national crisis induced by the blackout, continuity rather than transformation was the fate of Eskom as well. Initially, GEAR was to have entailed the privatization of Eskom (together with several other state holdings), but the ANC's union confederation ally Cosatu, together with their tripartite alliance partner, the SACP, succeeded in preventing the full unbundling of Eskom—under the debatable assumption that state ownership meant that they could rely on Eskom to maintain low electricity prices and otherwise act in the public interest, rather than on behalf of mining corporations and well-connected individuals.[49]

Perhaps the last opportunity during the Zuma presidency to fully mobilize the disruptive potential of RE toward transforming both South Africa's energy system and Eskom's governance came in May 2009, when National

---

[48] Mafeje (1978: 50–51). While this point of analytical departure stands in stark contrast to that of Mbeki's MA thesis on West African industry written a few years earlier, Mafeje—only a few years Mbeki's senior and from the same region of rural Eastern Cape province—was undoubtedly known to Mbeki at this time, as he had been a *cause célèbre* upon being refused an academic position at the University of Cape Town in 1968 on the grounds of race. He thereafter went into exile, returning to South Africa in 2000 and remaining until his death in 2007.

[49] PMG (2001).

Energy Regulator Nersa approved the Renewable energy Feed-in-Tariff (REFIT) policy to generate 10,000 GWh by 2013.[50] REFIT's proposed tariffs were initially designed to cover generation costs plus a real after-tax return on equity of 17%, fully indexed for inflation, for twenty years. In August 2009, the government devised the Electricity Regulations on New Generation Capacity ("New Gen Regs") as a green light to potential RE independent power producers. The Regulations demarcated clearly separate functions for the DoE, as the procurer of electricity from RE IPPs; described the buyer as an independent system and market operator (ISMO); and limited the power of Eskom to that of buyer and signatory of a power purchase agreement (PPA) independently approved by the regulator. Had the 2011 Independent Systems and Market Operator Bill been adopted at this time, the unbundling of Eskom would likely have been consolidated, and RE generation would soon thereafter have provided lower retail electricity prices as its infrastructure scaled up.

In early 2011, however, inter-ministerial rivalry and incoherence led to a flurry of contradictory statements that did not bode well for the REFIT scheme. First, Nersa announced its intention to review the tariffs set in March 2009, considering them to have been likely set too high, even though no power had yet been purchased at these initial rates. It missed its own mid-June deadline to announce the new rates, sending a cold-feet signal to potential investors. It then reduced the initial proposed rates by 25% for wind, 13% for concentrated solar, and 41% for photovoltaic—arguing that this was necessitated by "changes in exchange rates and the cost of debt". The new tariffs' capital component, moreover, would no longer be fully indexed for inflation. However, Energy Minister Dipuo Peters indicated that the 2009 REFIT rates, and not the revised 2011 tariffs, would be used for the first procurement round, and that the first 1000 MW would be procured by December 2011.[51]

By this time, the political utility of government tenders for coal procurement had grown. The recent reorganisation of government departments may have further contributed to confusion over their respective competencies. A National Treasury spokesperson claimed at the time that according to the Electricity Regulations Act, Nersa cannot "pre-determine a tariff" but only consider applications at the licensing stage on a "project

---

[50]DoE (2003).

[51]Creamer (2011).

application by project application" basis. Nersa regulatory member Thembani Bukula insisted however that it had secured legal opinion confirming that it was indeed authorised to determine tariffs, and that Nersa was taking its lead from the primary policymaker, the DoE, while ensuring alignment among the regulator, the National Treasury and the DoE. But despite the National Treasury insisting that "there is no misalignment with the Nersa process" and that there would be only one procurement process, which would be determined by Peters through Section 34 of the Electricity Regulations Act, uncertainty over the rules of engagement had by then increased.[52]

More damagingly, rather than stipulate conditions for participation and accept all comers who could meet them (the quickest way to increase RE capacity), it introduced new levels of ministerial oversight to determine eligibility continuously and establish which projects would be prioritized.[53] After the DoE and Treasury commissioned a legal opinion that argued, implausibly, that feed-in tariffs amounted to non-competitive procurement, it jettisoned the REFIT initiative. Then, after "informal feedback from the private sector on design, legal, and technology issues," the REIPPP Programme was launched.[54]

What had transpired between 2009 and 2011 to increase the government's inclination to jettison the REFIT in favour of the REIPPPP? Certainly, the decision to proceed with the Medupi and Kusile plants increased incentives to subordinate, rather than prioritize, RE generation. This also holds for further investments in RE storage. The cheapest and most widely available of these, pumped storage schemes for providing peaking generation, had enjoyed advanced research and planning by 2009. Eskom had identified a suite of viable sites between 1985 and 1995, and begun applying for permits from Nersa for schemes such as Tubatse pumped storage (planned at 1500 MW capacity) to add to the existing pumped storage capacity of almost 3000 MW. After Medupi got the go-ahead, however, the 2016 Draft Revision of the IRP 2010 made no mention of pumped storage capacity to be developed before 2030.[55]

---

[52] Creamer (2011).
[53] DoE (2011).
[54] Eberhard et al. (2014: 8).
[55] Barta (2018).

A less appreciated, but nonetheless highly significant, change transpiring shortly after Zuma assumed office was the enforcement of the new order rights under the 2002 Mineral and Petroleum Resources Development Act (MPRDA), which went into effect on April 30, 2009. Whereas the old common law-based 1991 Minerals Act gave land owners absolute rights over minerals found on or under their land, these are neutralised under the MPRDA, which regulates the acquisition of new order rights using BEE ownership and employment stipulations. It also makes transitional provisions for the continuation of old order mining rights, for which prior owners needed to apply within five years of the commencement of the MPRDA; failing which the old order mining right would cease to exist on April 30, 2009, unless this time period were extended by the Department of Minerals and Energy.

Several prior owners found that they would not make this deadline since they did not have adequate empowerment deals in place. Finding politically well-connected "assistance" thus became a top priority for them, particularly in the lucrative coal mining sector. The risk of conflicts of interest among public officials thereby grew commensurately.[56] At the same time, incentives for maintaining Eskom's relative autonomy in contracting coal as a tool of patronage and corruption—and coal's central importance, by deterring potentially competing RE sources that were less amenable to these ends—also increased. Thus, Zuma's government approved a third mega-coal station, Coal 3, in August 2013 and accelerated preparatory work for it the following year. The total amount allocated to BEE companies for inputs procurement from 2009 to 2014 increased sixfold from R20.8 billion to R125.4 billion—more than enough to pay for fifty more Sere Farm-type installations.

Conclusion

The rise of corrupt "tenderpreneurship" characterizing the Zuma presidency should be understood not as a challenge to, but on the contrary, as an extension of the power and control of the large mining multinationals over the sector and the economy more broadly.[57] The collusion that Eskom

---

[56] Among many examples, Mark Pamensky, the Chief Financial Officer of Oakbay Resources (among the companies organized and controlled by the Gupta brothers) at the time, was also the controlling shareholder of Tegeta (a Gupta-owned mine) while serving on the Eskom board as Investment and Finance Committee chair (Hofstatter 2018: 243–244).

[57] Desai et al. (2011).

CEO Brian Molefe exercised in 2016 by arranging R2.3 billion in forward purchase agreements to the Guptas to buy the Optimum coal mine on terms artificially disadvantageous to Glencore—shortly after Zuma's son Duduzane became a shareholder in the Gupta mining company Tegeta— seems extraordinary in its shamelessness and scale.[58] But it is nonetheless consonant with the broader pattern of interpenetration of Eskom and the large mining corporates.

The prior history of Optimum Coal Holdings illustrates this pattern. In June 2012, Lexshell 849 (owned by Cyril Ramaphosa, also Glencore's BEE partner) and a Glencore subsidiary bought 71% of Optimum Coal Holdings shares from a BEE consortium headed by Eliphus Monkoe (chief operating officer of BHP Billiton's coal division) and Mike Teke (BHP Billiton's shareholder director). The consortium, in turn, had bought the mine from BHP Billiton in 2008.[59] Similarly, BHP Billiton Energy Coal and Total Coal SA between them own 71% of Eloff Mining, with the balance owned by Canyon Springs, itself mostly owned by Siyanda Resources but controlled by Exxaro. Among BEE companies, only Exxaro, Shanduka (formerly owned by Ramaphosa prior to his becoming vice president in 2014) and Mmakau Mining (with dominant shareholder Brigette Radebe, wife of Justice Minister Jeff Radebe) have been in a legal and practical position to bid for Eskom's biggest contracts.[60] The small group of elites able to benefit from Eskom-related BEE tendering are thus at one and the same time major government players and substantial shareholders in the mining sector's largest firms.

More fundamentally, these transactions underscore the continuing, even intensifying, centrality of MEC activities that strive to deepen the country's economic dependence on coal. Related transactions include infrastructural investments that cumulatively cost over a trillion rand. These include the $53-billion rail line designed to export eighteen billion tons of coal on South China Rail locomotives (entailing $1 billion in kickbacks to the Gupta brothers); the Medupi and Kusile coal-fired power stations costing $15 billion each; and the $17 billion Durban port-petrochemical expansion, which aims to increase annual container traffic from 2.5 million to 20 million by 2040, and thereby facilitate coal

---

[58] Hofstatter (2018: 1–5, 227–228).

[59] Jaglin and Dubresson (2016: 143–145) and Hofstatter (2018: 123).

[60] Jaglin and Dubresson (2016: 146).

exports and accommodate prospective offshore oil and gas drilling by Sasol and other oil multinationals.[61]

Given the extent of the further interpenetration of the biggest BEE players and numerous members of the ANC's National Executive Committee, so long as the ANC remains South Africa's dominant political party, it is unlikely to instigate a major change in the MEC's basic structure, in which Eskom and coal mining play an essential part. This begs the question of which actors could achieve this end. As noted at the beginning of this chapter, struggles have changed state structures and social norms throughout South Africa's existence. Expanding on this theme, Cottle observes that the 1922 Rand Rebellion gave rise to a regime that protected white labour, the African Mine Workers' Strike of 1946 induced a reconfiguration of national politics and the establishment of apartheid, the 1973 Durban Strikes led to the reconfiguration of the industrial relations system and the emergence of an independent workers' movement, and the 2012 Marikana Strike precipitated a breakdown in the hegemony of the ANC and the establishment of new political opposition with the formation of the Economic Freedom Fighters (EFF) and the South African Federation of Trade Unions (SAFTU).[62] If Marikana becomes a true watershed in postapartheid history, this would imply a durable constraint on the state's recourse to such a naked use of force. The accelerating corruption in its wake is expressive of a regime with few other non-transformative options. "Between consent and force stands corruption, … characteristic of certain situations when it is hard to exercise the hegemonic function, and when the use of force is too risky."[63] A major question in this context, explored in subsequent chapters, is whether the major disruptive role in shifting the dominant energy mix and Eskom's *modus operandi* in it will be played by strands of the new opposition (or some similar coalition), by the ongoing rollout of large-scale RE provision, or some combination of both.

---

[61] Bond (2019).

[62] The EFF grew out of a faction of the ANC Youth League, and SAFTU is a labour federation that split off from the major confederation, Cosatu (Cottle 2019: 45).

[63] Gramsci (1992 [1930–1931], Notebook 1, §37: 156).

References

Barta, B. (2018, March 9). The contribution of pumped storage schemes to energy generation in South Africa. *Creamer Media's Engineering News*. Accessible at: https://www.ee.co.za/article/the-contribution-of-pumped-storage-schemes-to-energy-generation-in-south-africa.html.

Basson, A., & Du Toit, P. (2017). *Enemy of the people*. Cape Town: Jonathan Ball.

Bhorat, H., & Swilling, M. (2017). *Betrayal of the promise: How South Africa is being stolen* (State Capacity Research Project).

Bond, P. (2019, January 1). South Africa suffers capitalist crisis Déjà Vu. *Monthly Review*. Accessible at: https://monthlyreview.org/2019/01/01/south-africa-suffers-capitalist-crisis-deja-vu/.

Bratsis, P. (2014). Political corruption in the age of transnational capitalism. *Historical Materialism, 22*(1), 105–128.

CDE. (2008, July). *South Africa's electricity crisis: How did we get here? And how do we put things right?* (CDE Roundtable No. 10). Johannesburg: Centre for Development and Enterprise.

Christie, R. (1984). *Electricity, industry and class in South Africa*. London: Palgrave Macmillan.

Cottle, E. (2019). Competing Marxist Theories on the Temporal Aspects of Strike Waves: Silver's Product Cycle Theory and Mandel's Long Wave Theory. *Global Labour Journal, 10(1)*, 37–50.

Creamer, T. (2011, June 23). Fresh concern that SA is poised to abandon REFIT in favour of competitive bidding. *Creamer Media's Engineering News*. Accessible at: http://www.engineeringnews.co.za/article/fresh-concern-that-SA-is-poised-to-abandon-reFIT-in-favour-of-competitive-bidding-2011-06-23.

Department of Energy (DoE). (2009). *Creating an enabling environment for distributed power generation in the South African electricity supply industry*. Pretoria: DoE.

Desai, A., Maharaj, B., & Bond, P. (2011). Introduction: Poverty eradication as holy grail. In B. Maharaj, A. Desai, & Bond, P. (Eds.), *Zuma's Own Goal: Losing South Africa's "War on Poverty"* (pp. 1–35). Trenton, NJ: Africa World Press.

DME. (1998). *White paper on the energy policy of the Republic of South Africa*. Department of Minerals and Energy, Government of South Africa. Accessible at: http://www.energy.gov.za/files/policies/whitepaper_energypolicy_1998.pdf.

DME. (2003). *White paper on renewable energy*. Department of Minerals and Energy, Government of South Africa. Accessible at: https://unfccc.int/files/meetings/seminar/application/pdf/sem_sup1_south_africa.pdf.

Department of Energy (DoE). (2003). *South African Department of Energy, White Paper on Renewable Energy*. Accessible at: https://www.energy.gov.za/files/policies/whitepaper_renewables_2003.pdf.

Department of Energy (DoE). (2011). *Electricity Regulations on New Generation Capacity.* Accessible at: https://www.energy.gov.za/files/policies/Electricity%20Regulations%20on%20New%20Generation%20Capacity%201-34262%204-5.pdf.

Donnelly, L. (2009, November 6). Eskom upheaval: The case against Maroga. *Mail & Guardian.* Accessible at: https://mg.co.za/article/2009-11-06-eskom-upheaval-the-case-against-maroga.

Eberhard, A. (2005). From state to market and back again: South Africa's power sector reforms. *Economic and Political Weekly, 40*(50), 5309–5317.

Ebarhard, A. (2007). The political economy of power sector reform in South Africa. In D. Victor & T. Heller (Eds.), *The Political Economy of Power Sector Reform* (pp. 215–253). New York: Cambridge University Press.

Eberhard, A., Kolker, J., & Leigland, J. (2014). *South Africa's renewable energy IPP procurement program: Success factors and lessons.* Accessible at: https://www.gsb.uct.ac.za/files/ppiafreport.pdf.

Ellis, S. (2013). *External mission: The ANC in exile, 1960–1990.* Oxford: Oxford University Press.

ENCA. (2013, March 22). Eskom-BHP contract a 'scandal'. *ENCA.* Accessible at: https://www.enca.com/money/eskom-bhp-contract-scandal.

Feinstein, A. (2007). *After the party: A personal and political journey inside the ANC.* Cape Town: Jonathan Ball.

Gramsci, A. (1992). *Prison notebooks* (J. Buttigieg, Trans., Vol. 1). New York: Columbia University Press.

Habib, A. (2013). *South Africa's Suspended Revolution: Hopes and Prospects.* Johannesburg: Wits University Press.

Hofstatter, S. (2018). *Licence to loot: How the plunder of Eskom and other parastatals almost sank South Africa.* Cape Town: Penguin Random House South Africa.

Huntington, S. P. (1968). *Political order in changing societies.* New Haven: Yale University Press.

Jaglin, S., & Dubresson, A. (2016). *ESKOM: Electricity and Technopolitics in South Africa.* Cape Town: Juta and Company (Pty) Ltd.

Kasrils, R. (2018). *Armed and dangerous: From undercover struggle to freedom* (4th ed.). Johannesburg: Jacana Media.

Le Roux, M. (2006, March 30). Mbeki: There is no electricity crisis. *Mail & Guardian.* Accessible at: https://mg.co.za/article/2006-03-30-mbeki-there-is-no-electricity-crisis.

Lodge, T. (2014, January 1). Neo-patrimonial politics in the ANC. *African Affairs, 113*(450), 1–23. https://doi.org/10.1093/afraf/adt069.

Mafeje, A. (1978). *Science, ideology, and development: Three essays on development theory.* Uppsala: Scandinavian Institute of Development Studies.

Malala, J. (2015). *We have now begun our descent: How to stop South Africa losing its way.* Cape Town: Jonathan Ball.

Marais, H. (2011). *South Africa pushed to the limit: The political economy of change.* London: Zed Books.

MPE. (2000). *An accelerated agenda toward the restructuring of state owned enterprises: Policy framework.* Pretoria: Ministry of Public Enterprises. Accessible at: https://www.gov.za/sites/default/files/gcis_document/201409/acceleratedagendarestructuringsoe0.pdf.

Padayachee, V. (1991). The politics of South Africa's international financial relations, 1970–1990. In S. Gelb (Ed.), *South Africa's economic crisis.* Cape Town: David Philip.

Pauw, J. (2017). *The president's keepers.* Cape Town: Tafelberg.

PFMA. (1999). *Public Finance Management Act of South Africa.* Accessible at: http://saflii.austlii.edu.au/za/legis/hist_act/pfma1o1999225/pfma1o1999a2s2013356.html.

PMG. (2001, June 11). *Eskom Conversion Bill: Cosatu input; Department briefing on additional amendments.* Accessible at: https://pmg.org.za/committee-meeting/596/.

Satgar, V. (Ed.). (2018). *The Climate Crisis: South African and Global Democratic Eco-Socialist Alternatives.* Johannesburg: Wits University Press.

Styan, J. (2015). *Blackout: The Eskom crisis.* Cape Town: Jonathan Ball.

Van Vuuren, H. (2006, May). *Apartheid grand corruption: Assessing the scale of crimes of profit from 1976 to 1994.* A report prepared by civil society in terms of a resolution of the Second National Anti-corruption Summit for presentation at the National Anti-corruption Forum. Cape Town: Corruption and Governance Institute for Security Studies.

Yelland, C. (2009). *Independent Power Producers (IPPs) Organise Collectively to Take on Eskom.* EE Publishers.

Yelland, C. (2011, April 28). Eskom, BHP Billiton and the secret electricity pricing deals. *Daily Maverick.* Accessible at: https://www.dailymaverick.co.za/article/2011-04-28-eskom-bhp-billiton-and-the-secret-electricity-pricing-deals/.

CHAPTER 4

Non-RE Alternative Energies: Nuclear, Geothermal, Fracking and Offshore Gas

Abstract Despite obvious differences, South Africa's main contending sources of non-renewable energy apart from coal—nuclear, geothermal, and (fracked or offshore) natural gas—all share three attributes. They are all capital intensive, requiring huge up-front investments in order to break even (thus also requiring an active state role in mitigating risk); they entail significant, probably irredeemably large, externalities and environmental costs that magnify, by orders of magnitude, risks of intergenerational injustice; and related to these, they cannot effectively address South Africa's historic energy injustices, nor improve its level of energy security. Of these three, South Africa has extensive experience only with nuclear power: Its co-evolutionary relationship with the apartheid state and related culture of secrecy and lack of accountability should serve as an object lesson regarding the development of the other sources discussed.

Keywords Nuclear energy · Geothermal energy · Fracking · Offshore gas · Intergenerational injustice

This chapter discusses three energy sources that—despite several obvious differences—share three attributes: they are all capital intensive, requiring huge up-front investments in order to break even (thus also requiring an active state role in mitigating risk); they entail significant, probably irredeemably large, externalities and environmental costs that magnify, by

A. Lawrence, *South Africa's Energy Transition*, Progressive Energy Policy, https://doi.org/10.1007/978-3-030-18903-7_4

orders of magnitude, risks of intergenerational injustice; and related to these, they cannot effectively address South Africa's historic energy injustices, nor improve its level of energy security.

GEOTHERMAL ENERGY, FRACKING, AND OFFSHORE GAS

Because of its lack of major volcanic activity, South Africa's geothermal energy potential had been disregarded until recently. A recent study by the Council of Geoscience suggests, however, that five regions—in the Western Cape, Karoo, Northwest, Limpopo, and northern KwaZulu-Natal provinces—show significant volumes of buried heat accessible via low-enthalpy geothermal energy harvesting. However, given the technology's similarities with fracking and attendant water table pollution risks, and the fact that the source of heat is often the same radioactive ores that the mining industry has left strewn among watersheds across the country, the technology has more risks and uncertainties associated with it—and higher costs—than current wind and solar deployments do. A geothermal feasibility test at a $38 million pilot plant in Pohang, South Korea in November 2017 was responsible for a magnitude 5.4 earthquake—that county's second worst ever recorded—causing $54 million in damages, and demonstrating that the technique's risks were far greater than anticipated.[1] Unlike fracking and nuclear power (notwithstanding the close mining industry ties of all three), however, geothermal has no significant active lobby pushing it as an energy alternative.[2]

Fracking similarly combines clear risks with uncertain rewards. Lobbying in its favour by local and transnational oil interests has escalated over the past decade. In the face of widespread public protest, environmental consulting firm SLR Consulting released a report in September 2016 recommending that Rhino Oil and Gas Exploration's exploration application be granted for 1,500,000 ha (15,000 km2) of KwaZulu Natal's central Midlands traversing nearly 10,000 farms.[3] The following year, in March 2017, Mineral Resources Minister Mosebenzi Zwane announced that it would grant exploratory permits to three corporations for exploration of three different areas within the Karoo that together make up a fifth of South

---

[1] Voosen (2018).

[2] Dhansay et al. (2017).

[3] https://www.news24.com/SouthAfrica/News/no-fracking-victory-20160919.

Africa's total land area: Shell SA for 90,000 km^2, Falcon Gas and Oil for 32,000 km^2, and Challenger Energy/Bundu Oil and Gas for 3200 km^2. The government rationale was that shale gas would be cost competitive with coal and "significantly reduce [South Africa's] carbon footprint".[4]

The announcement was made in the face of growing community complaints about a lack of consultation and access to information. As with Eskom's relationship with the coal industry, the regulatory framework for fracking contains clear conflicts of interest. The oil and gas regulator, the Petroleum Agency of South Africa, while officially promoting the oil and gas industry, is at the same time the entity responsible for granting exploration rights permits. Concerning the Shell SA exploration award, it is no coincidence that the metonym for the ANC National Executive is Shell House (its building headquarters in Johannesburg), as Shell was the original donor of the funds for the building. More recently, Shell SA's gas distribution operation is 25% owned by Thebe Investment Corporation, which in turn is almost 50% owned by Batho Batho Trust, an ANC investment company.[5]

Ostensibly, the Mineral and Petroleum Resources Development Act (MPRDA) governs applications for and the granting of rights and permits to conduct shale gas extraction, which requires clearance from the 1998 National Environmental Management Water Act (NEMWA), the framework for governing Section 24 of the Constitution that stipulates the right of every person to a non-harmful environment while simultaneously mandating the government to protect the environment. NEMWA would then determine whether the proposed activity raises environmental protection concerns and demands, if required, that any requisite waste management licenses be applied for and obtained prior to starting any activity requiring the management of the waste. Should any company succeed with exploration, a water use license in terms of the National Water Act is to be secured through applications for individual or integrated water use licenses. Once a permit applicant publishes its Environmental Management Plan, members of the public who may be affected parties have 30 days to comment.

Given the abject failure of similar regulations around the much better-known mining industry, and its widespread abrogation of water licence requirements, local communities and civil society groups had little faith that

---

[4] Etheridge (2017).
[5] Fig (2018: 260).

this regulatory structure would be sufficient to fully consider the potential risks entailed. Responding to the threat of litigation around the imperfections of this process, the (then) Minister of Mineral Resources, Susan Shabangu, declared a moratorium on the issuing of exploration licences in February 2011, only to lift the moratorium in September 2012, while mandating the Minister of the Department of Mineral Resources to hold a series of public consultations with interested and affected stakeholders to provide further details.

The consultancy McKinsey & Co. argues that "The urgency to act on the natural gas opportunity is growing"—without explaining why, or whence the urgency emanates. Rather than recommend completing environmental impact assessments first, prior to issuing permits, it encourages the government to "finalise regulations, issue permits for pilot wells, and simultaneously complete environmental impact assessments. It could also guarantee purchase of the gas as an end-user for a number of years" in order to "support a low-carbon, cost-effective power sector and allow for greater energy independence … [and thereby] boost South Africa's GDP by 138 billion to 251 billion rand ($12 billion to $22 billion) by 2030 [and thus] create up to 328,000 direct and indirect jobs". At the same time, it admits that the "price of gas would need to drop below $6 per MMBtu by 2030 to begin to displace coal on the cost curve…. We acknowledge that there is uncertainty about the viability of shale and offshore gas resources and that the long-term role of natural gas in South Africa will not be clear for many years."[6]

One dimension of uncertainty they fail to acknowledge in their report is that of the extent of environmental damage. The likelihood of water contamination from large-scale fracking has been shown to be a near-certainty from the experience of much better-regulated national environments such as that of the United States.[7] Fracking requires pumping millions of litres of water, mixed with toxic chemicals and sand, at high pressure into underground shale rock formations. The principal location of fracking interest is the Karoo, a semi-arid to arid expanse straddling the Western Cape and Northern Cape provinces—thus an area already severely water stressed.

Hazards to the regional environment include refluxed or accidentally spilled fracking liquid wastewater disposal (which typically consists of both

[6]McKinsey (2015).

[7]See i.a. Howarth et al. (2011) and Mooney (2011).

toxic and radioactive sludge) as well as from released natural gas polluting local water tables, and thereby contaminating drinking-water. Studies have identified health risks to communities living close to fracking sites including increased rates of cancer, liver damage, immunodeficiency, and neurological symptoms; increased risk of birth defects, asthma and other respiratory ailments, and occupational health and safety problems like exposure to crystalline silica, a type of sand used during fracking. Other hazards include adverse impacts to soil, air, and water quality in drilling regions; increased incidence of earthquakes up to 10 km from a given fracking site; and increases in volatile organic compounds and air toxics locally as significant health threats.[8]

Discourse about fracking as "low carbon" and helping "reduce South Africa's carbon footprint" are particularly disingenuous given that the risk of methane leakage (a greenhouse gas more than 80 times more potent than CO^2) is well established. One recent leak at Aliso Canyon near Los Angeles, CA in 2016 released more than 100,000 tons of methane: as much pollution as 600,000 cars emit in one year. Although such leakage can be reduced using reduced emissions completions, given that the far more straightforward requirements of particulate removal were stipulated but never enacted for the Medupi and Kusile coal generation plants, there is little prospect that such regulatory measures would be adopted for South African fracking.[9]

The same cautionary argument applies for offshore deepwater natural gas exploration. The Italian fossil fuel conglomerate ENI, in conjunction with Sasol, already has obtained exploration and development rights off the northern coast of Mozambique, and together they seek to gain similar extractive rights off the coast of KwaZulu Natal between St. Lucia and Port Shepstone, east of Durban harbour, having gained an exploratory drilling permit to commence in 2019. The project is one of many that Sasol and ENI plan under the auspices of "Operation Phakisa," which aims to use the diminished regulatory clout of South Africa's Exclusive Economic Zone to locate and drill a target of thirty exploration wells before 2024. A coalition of community organizations and environmental groups (including WildOceans, Wild Trust, South Durban Community Environmental Alliance, GroundWork, and others) actively oppose the plan, citing

---

[8]Busby and Mangano (2017), Esswein et al. (2013), Magnani et al. (2017), Willis et al. (2018), and Johnston et al. (2019).

[9]Umeozor et al. (2018).

endangerment to wildlife—such as the critically endangered Coelacanth, humpback whale, Loggerhead and Leatherback turtle populations, and sardine shoals—no less than to the local tourism and fishing industries.[10] In a packed South Durban community meeting in May 2018, the diverse, mostly working class civil society coalition present expressed outrage at the Sasol representative present for suggesting the plan would benefit the local economy, arguing that, on the contrary, it would destroy the livelihoods and habitat of precisely the most vulnerable members of the community, for generations to come, and promised further legal action against the plan.[11] In addition to environmental risk hazards and resource curse effects, South Africa's offshore gas faces a dilemma similar to that of its coal sector: the global markets confront a looming glut in production, driven by the US shale revolution. It is questionable whether any such finds could be developed at scale at depressed global prices.

Nuclear Energy: Isihloko Asivaliwe?

Perhaps no other major energy source better illustrates the role of transnational interests and networks in its establishment and further promotion than nuclear energy. The sector is unusual in some respects: South Africa's is the only nuclear exemplar in Africa, remains one of the few (together with Canada) capable of sourcing its uranium domestically, and remains the only case of a country that unilaterally and voluntarily dismantled its own nuclear weapons program. But in most other respects, it is typical of other nuclear power contexts. Power plant construction is prone to massive delays; budget limits are routinely and massively exceeded; day-to-day operations are shrouded in secrecy. As with coal generation, nuclear's costs per kWh—even excluding those of decommission and waste storage—increasingly outstrip those for wind and solar power.[12] Unlike these RE sources, however, South African nuclear power enjoys the advantage of a well-established lobby that has managed to keep the sector alive politically against seemingly insuperable economic and practical odds.

On the eve of the election of the National Party government in 1948 that would fatefully enact officially proclaimed apartheid legislation, then-Prime Minister Jan Smuts was contacted by his UK former counterpart (now

---

[10] Pieterse (2018).

[11] Author's observations; Nxumalo (2018).

[12] Lazard (2018), cf. fn 9, above.

Opposition leader), Winston Churchill, regarding the potential sourcing of uranium to produce nuclear weapons, then in short supply. Smuts informed Churchill that from the 1920s onwards, South African geologists had detected substantial quantities of uranium as a by-product of gold mining—since the two minerals were often found together in the same mined ore. This prompted the UK and US to engage South Africa in a secret agreement over the next two decades, whereby they would finance the mining and purchase all available uranium at a premium. Smuts facilitated this process through the creation of a secret Atomic Energy Board (AEB), kept separate from the Council of Scientific and Industrial Research (CSIR) and answerable only to him, that would supervise all nuclear-related research activities.[13]

The immediate effect was to give the gold industry, and Anglo American Corporation in particular, a badly needed shot in the arm, since it had been experiencing a crisis of profitability compounded by wartime restrictions and growing labour militancy. The longer-term effect was to strengthen the opaque exercise of power of whichever regime was in power: first, the apartheid regime—which gave the AEB headquarters and first reactor site the honest name of "Pelindaba," meaning "no more talking" in Zulu—and now, its post-apartheid successor. Soon thereafter, in 1957, the South African government signed a secret agreement of nuclear cooperation with the United States, the United States of America and Union of South Africa Agreement for Co-operation Concerning the Civil Uses of Atomic Energy, allowing the purchase of weapons grade uranium from the United States.[14]

By the 1960s, to nuclear South Africa's initial utility to the West as reliable uranium supplier was augmented its roles as profitable market for Western multinational firms' products and foreign direct investment, and regional hegemon against the putative Communist threat posed by excessively independent movements of national independence. South Africa's industry expanded by establishing its own uranium enrichment plant (at a facility next to Pelindaba, "Valindaba," meaning "the subject is closed"), ostensibly for civilian purposes, but actually to produce and assemble its own fleet of weapons, whose existence was publicly acknowledged for the

---

[13]Fig (2005: 38–39).
[14]UN (1957).

first time by then-President de Klerk, and whose total costs may have reached R5 billion by that time.[15]

Despite the findings of the Forsyth Commission of Inquiry in 1961 that a nuclear power plant in Cape Town or anywhere else in South Africa would never make economic sense, security interests had prevailed, leading to the siting of a new plant at Koeberg, suburban Cape Town, in the 1970s. The nuclear industry would thereby outlive its military-use origins and persist into the democratic era. The advent of political liberalisation in 1990, including the unbanning of political parties and release of political prisoners, heralded the likely victory of the ANC. The apartheid government therefore decided to dismantle its nuclear weapons program, including the six bombs already built, and future weapons capacity, and became signatory to the Nuclear Non-proliferation Treaty in 1991.

Built with the assistance of a French-led consortium, Koeberg's total cost estimates predictably ballooned from R600 million in 1975 to over R1.8 billion in 1984, when the first of two reactors began operation, not including a further R68 million in repairs needed after anti-apartheid activists succeeded in damaging switching equipment with a bomb in 1982. It thereby inadvertently helped to hasten the defeat of apartheid by adding to the mounting fiscal crisis that the state had incurred. Effective regulatory oversight remained a challenge after democratization, however. The government's Council for Nuclear Safety, established in 1988, struggled to attain sufficient independence from industry interests, and was replaced under the 1999 National Nuclear Regulatory Act with a newly established National Nuclear Regulator (NNR), and the Nuclear Energy Committee of South Africa (NECSA) replacing the AEC. Its board was to include stakeholder representatives from the community, business, and organised labour, although these positions also went unfilled for many years. Similarly inadequate regulatory oversight has plagued the nuclear waste storage site of Vaalputs, where intermediate-level radioactive waste leakage was discovered in 1997.[16]

Despite the new government's decision to save money through terminating its enrichment facilities, it persisted in funding the unproven—and unviable—pebble bed nuclear reactor model. Work on the Pebble Bed Modular Reactor was started in South Africa in 1992 and made public

[15]Fig (2005: 42–45).
[16]Fig (2005: 56–63).

in 1998 when it was forecast that commercial reactors would be available for purchase from 2004 onwards. Development was officially abandoned in 2010 (and unofficially by 2004) after expenditure of almost R9 billion of public money, with a forecast requirement of R31 billion, when no orders were in sight. The design was not sufficiently developed to submit it to the South African safety regulator, the NNR; a daunting task, as the NNR would be the first regulator to review a new and innovative design. Eskom estimated commercial units would not be available before about 2030 and forecast costs per reactor had escalated many-fold.[17]

Yet South African and international nuclear lobby interests had not given up on more conventional nuclear reactor models to expand upon Koeberg's capacity, and in fact the 2010 IRP included provision for increased nuclear capacity of 9600 MW. In 2015, then-Finance Minister Nhlanhla Nene granted the DoE R200 million with which to research the procurement process; soon thereafter, Cabinet approved the procurement initiative, albeit using the dated but still current version of the IRP, and thus failing to acknowledge that South Africa's required energy projections had decreased. In 2011, Cabinet inaugurated a National Nuclear Energy Executive Procurement Coordinating Committee, reassigned to the presidency in 2013 and expanded in 2014, when the National Radioactive Waste Disposal Institute (NRWDI) was launched. Barely a year later, however, the minister of energy set up a task team to investigate serious mismanagement at both NRWDI and NECSA, which had a deficit of nearly R150 million and whose required report was more than two years overdue.[18]

It remains unclear what aspect of the procurement process needed R200 million-worth of further research, however, since a year *before* these funds were granted, a secret deal had already been reached between Zuma and Russia's President Putin. Unlike prior memoranda of understanding with other potential nuclear vendors, this one contained legally binding contractual details, such as South Africa's ceding full control of its domestic nuclear industry to Russia; Russia's contract would be tax-exempt; and Russia would be indemnified against any liabilities for nuclear damage or accident, including those occurring outside of the country.[19] The lack of regulatory oversight exercised up to this point would seem to increase the

[17] Thomas (2009) and Fig (2005: 64).

[18] Fig (2018: 258–261).

[19] Gosling (2017).

likelihood of accidents, whose chances of severity are increased by the proposed plant sites' proximity to some of the most densely populated areas of the country.[20]

Things finally came to a head in December 2015, when then-Finance Minister Nene insisted that South Africa could not afford a R1 trillion + nuclear procurement deal, and was promptly fired by Zuma (who had obtained Cabinet approval for the nuclear programme) and replaced with a yes-man utterly lacking in qualifications for the post. The financial markets responded immediately, with the rand losing 10% of its value and the downgrading of South Africa's bonds to junk status suddenly becoming a real threat. Finally, on 26 April 2017 (the thirty-first anniversary of the Chernobyl disaster) the Western Cape High Court ruled in favour of litigants EarthLife Africa and the Southern African Faith Communities' Environmental Institute that the procurement was illegal for having failed to follow due process; for being placed under Eskom's auspices; and (together with the other nuclear memoranda of understanding) for having failed to come before parliament to be discussed openly. Similarly, Nersa was found to be derelict for rubber-stamping the deal instead of ensuring public participation.[21] The following year, newly appointed Energy Minister Jeff Radebe sacked the previous NECSA board and suspended CEO Phumzile Tshelane, ex-board chairperson Kelvin Kemm, and former audit and compliance board member Pamela Bosman, arguing insubordination, failure of legislative compliance and various irregularities that cumulatively have cost NECSA and its subsidiaries losses totalling over R250 million.

What if these procedures had been followed as scrupulously as the High Court ruled is necessary? The power to approve nuclear power plant procurements rests with parliament, and their inclination to do so depends on presidential leadership and popular demands. The ANC's parliamentary majority is unlikely to oppose the president's wishes, except in the most extraordinary circumstances.[22] Large-scale survey data on South Africans' opinions about nuclear power is non-existent, but a recent small survey suggests a majority are in favour.[23] At the start of his term, President

---

[20] See e.g. Carnie (2015).

[21] Fig (2018: 267–268).

[22] However, in August 2017, when the speaker of the national assembly, Baleka Mbete, exercised his constitutional authority to announce a secret ballot on a motion of no confidence in then-President Zuma, it was only narrowly defeated, with 198 against v. 177 in favour.

[23] Nkosi and Dikgang (2018).

Ramaphosa appears unlikely to recommend nuclear builds. At the tenth BRICS Summit in Johannesburg in July 2018, he cited severe financial constraints as the chief reason against them; the following month, Energy Minister Jeff Radebe reiterated this view, observing that aggregate demand had fallen to its 2007 level and stating that said the new Cabinet had shelved Zuma's nuclear plans and would opt for using coal, gas and wind power instead.[24]

Yet elite-level consensus on this point may not prove very durable. South Africa will discuss nuclear energy "cooperation" with Russia during the next high-level bilateral meeting, Lindiwe Sisulu, the South African minister of international relations and cooperation, was quoted as confirming. "I think our president [Cyril Ramaphosa] was opting for a mixed use for nuclear energy and I think [he and Russian President Vladimir Putin at the last meeting] were in complete agreement about that, but it was the timing that was a bit too soon for us to be able to absorb that. I am not sure when we will be going to Russia, but I am certain when we go to Russia the matter will be on the table."[25]

In light of such comments, it isn't too soon for the South African public to absorb that nuclear energy will never be the lowest-cost nor the cleanest energy source currently available. The survey results also show that favourable attitudes to nuclear power are likely to diminish as perceptions of risk and danger, or increased electricity costs, increase.[26] In a political context where government officials routinely refer to practically any energy source—such as fracking and offshore oil—as "low carbon," it is easy to forget that nuclear power's low-carbon credentials are similarly inflated. Not only is the level of emissions from all stages of the nuclear lifecycle (including uranium mining and milling, fuel processing, waste storage, and decommissioning) substantial, as plants are obliged to use increasingly low ore quality, their carbon footprint will continue to grow.[27] With very low ore grades in use, some nuclear power plants currently emit the equivalent of 337 g CO_2/kWh, making them *already* as environmentally damaging as equivalent-sized gas-fired power plants.[28] Moreover, the many years that

---

[24] Kings (2018).

[25] Rizvi (2019).

[26] Nkosi and Dikgang (2018).

[27] Sovacool (2008) and Mudd and Diesendorf (2008).

[28] Beerten et al. (2009).

planning and construction require mean that in even the rosiest scenarios, nuclear energy will not be available for South Africa when it needs to transition from coal most urgently, over the next five to ten years. The same is true for the other sources discussed—fracked or offshore gas and geothermal—none of which lends itself to addressing energy poverty either. It is evident in any case that Zuma's nuclear plans were intended precisely to complement, not displace, coal generation.

CONCLUSION

Given the commonalities among the major types of non-renewable alternatives to coal—their immense capital expenditure requirements and massive cost overruns; their uncertain technologies and risks; their potential for significant environmental damage lasting generations; the lack of transparency surrounding their acquisition, and related risks of accelerating corruption and capital flight; and their dubious (but precisely because they are so capital-intensive, likely very limited) employment creation potential—it may seem surprising that their fates should have been so different up to the present. Yet the difference in outcomes is easily explicable with reference to size of the anticipated corporate profits, and the extent to which the corporate interests are embedded within existing state structures. At one extreme, geothermal presents comparatively minor, uncertain profits with little embeddedness at present; the Russian (or any other) nuclear deal presented potentially sizeable profits, but little embeddedness (except within the small local nuclear lobby and ex-President Zuma's immediate circle); whereas fracking and offshore gas exploration combine potentially larger profits with higher levels of government embeddedness. In terms of total cost, employment outcomes, pollution, health hazards, and resource curse effects, however, the grounds for opposing each of these "alternatives" is fundamentally the same as for opposing the coal-based status quo.

REFERENCES

Beerten, J., Laes, G., Meskens, G., & D'haeseleer, W. (2009, December). Greenhouse gas emissions in the nuclear life cycle: A balanced appraisal. *Energy Policy, 37*(12), 5056–5068.

Busby, C., & Mangano, J. (2017). There's a world going on underground—Infant mortality and fracking in Pennsylvania. *Journal of Environmental Protection, 8,* 381–393. https://doi.org/10.4236/jep.2017.84028.

Carnie, T. (2015, July 16). At least one nuke power station for KZN. *IOL News*. Accessible at: http://www.iol.co.za/news/south-africa/kwazulu-natal/at-least-one-nuke-power-station-for-kzn-1886079.

Dhansay, T., Musekiwa, C., Ntholi, T., Chevallier, L., Cole, D., & de Wit, M. J. (2017). South Africa's geothermal energy hotspots inferred from subsurface temperature and geology. *South African Journal of Science*. http://doi.org/10.17159/sajs.2017/20170092.

Esswein, E., Breitenstein, M., Snawder, J., Kiefer, M., & Sieber, W. (2013). Occupational exposures to respirable crystalline silica during hydraulic fracturing. *Journal of Occupational and Environmental Hygiene, 10*(7), 347–356.

Etheridge, J. (2017, March 30). Government gives green light for shale gas fracking in Karoo. *News24*. Accessible at: https://www.news24.com/SouthAfrica/News/govt-gives-green-light-for-shale-gas-fracking-in-karoo-20170330?fbclid=IwAR0L1O_QIpA4KUCRWxH53EqZiFoxmgvN43RtEkHVy-qrUfwahtX8yhB1M18.

Fig, D. (2005). *Uranium road: Questioning South Africa's nuclear direction*. Cape Town: Jacana Media.

Fig, D. (2018). Capital, climate and the politics of nuclear procurement in South Africa. In V. Satgar (Ed.), *The climate crisis: South African and global democratic eco-socialist alternatives* (pp. 252–271). Johannesburg: Wits University Press.

Gosling, M. (2017, February 25). Russian nuclear deal places massive liability on South Africans. *News24*. Accessible at: https://www.news24.com/SouthAfrica/News/russian-nuclear-deal-places-massive-liability-on-south-africans-20170225.

Howarth, R. W., Ingraffea, A., & Engelder, T. (2011). Natural gas: Should fracking stop? *Nature, 477*(7364), 271.

Johnston, J., Lim, E., & Roh, H. (2019, March 20). Impact of upstream oil extraction and environmental public health: A review of the evidence. *Science of the Total Environment, 657*, 187–199.

Kings, S. (2018, August 27). South Africa has a new energy plan. *Mail & Guardian*. Accessible at: https://mg.co.za/article/2018-08-27-south-africa-has-a-new-energy-plan.

Lazard. (2018). *Levelized cost of energy data*. Accessible at: https://www.lazard.com/media/450784/lazards-levelized-cost-of-energy-version-120-vfinal.pdf.

Magnani, M. B., Blanpied, M. L., DeShon, H. R., & Hornbach, M. J. (2017). Discriminating between natural versus induced seismicity from long-term deformation history of intraplate faults. *Science Advances, 3*(11), e1701593.

McKinsey & Co. (2015).*South Africa's big five: Bold priorities for inclusive growth*. McKinsey Global Institute. Accessible at: https://studyres.com/doc/15857434/south-africa-s-big-five.

Mooney, C. (2011). The truth about fracking. *Scientific American, 305*(5), 80–85.

Mudd, G., & Diesendorf, M. (2008). Sustainability of uranium mining and milling: Toward quantifying resources and eco-efficiency. *Environmental Science & Technology, 42*(7), 2624–2629.

Nkosi, N., & Dikgang, J. (2018, January). *South African attitudes about nuclear power: The case of the nuclear energy expansion* (ERSA Working Paper No. 726). Pretoria. Accessible at: https://econrsa.org/2017/wp-content/uploads/working_paper_726.pdf.

Nxumalo, M. (2018, December 6). Environmental lobby prepares legal fight over offshore drilling. *Daily News.* Accessible at: https://www.iol.co.za/dailynews/environmental-lobby-prepares-legal-fight-over-offshore-drilling-18406392.

Pieterse, C. (2018, August 21). Oil threat to KZN coast. *The Witness.*

Rizvi, F. (2019, February 11). South Africa ready to discuss nuclear energy cooperation with Russia—Foreign Minister. *UrduPoint.* Accessible at: https://www.urdupoint.com/en/world/rpt-south-africa-ready-to-discuss-nuclear-e-549632.html.

Sovacool, B. (2008, August). Valuing the greenhouse gas emissions from nuclear power: A critical survey. *Energy Policy, 36*(8), 2940–2953.

Thomas, S. (2009, June 22). The demise of the Pebble Bed Modular Reactor. *Bulletin of the Atomic Scientists.*

Umeozor, E. C., Jordaan, S. M., & Gates, I. D. (2018). On methane emissions from shale gas development. *Energy, 152,* 594–600.

UN. (1957, July 8). *Treaty No. 4234 between the United States of America and the Union of South Africa.* Agreement for co-operation concerning the civil uses of atomic energy. Washington, DC. Accessible at: https://treaties.un.org/doc/Publication/UNTS/Volume%20290/volume-290-I-4234-English.pdf.

Voosen, P. (2018, April 26). Second-largest earthquake in modern South Korean history tied to geothermal plant. *Science.* Accessible at: http://www.sciencemag.org/news/2018/04/second-largest-earthquake-modern-south-korean-history-tied-geothermal-plant.

Willis, M., Jusko, T., Halterman, J., & Hill, E. (2018, October). Unconventional natural gas development and pediatric asthma hospitalizations in Pennsylvania. *Environmental Research, 166,* 402–408. https://doi.org/10.1016/j.envres.2018.06.022.

CHAPTER 5

REIPPPP: Renewables' Rise, or REIPPPP RIP?

Abstract South Africa's first and only sustained attempt to harness its renewable energy endowment, its Renewable Energy Independent Power Producers Procurement Programme (REIPPPP), has been heralded as a great success of inter-ministerial cooperation in attracting billions of rands of investment. However, it has failed to make substantial inroads in mitigating, let alone reversing, South Africa's high aggregate emissions levels and vicious cycle of poverty, inequality, and unemployment driven by the mining-centred economy. Its design has served to supplement, rather than replace, coal generation. Similarly, socioeconomic development goals of promoting localized industry, increasing employment, and ensuring the constitutional right to clean and affordable necessities such as energy access have been subordinated to that of creating an investment vehicle attractive to local and transnational financial interests. A realistic prospect of realising these goals may necessitate replacing the REIPPPP for an alternative means of promoting RE generation.

Keywords Renewable Energy Independent Power Producers Procurement Programme (REIPPPP) · Black Economic Empowerment (BEE) · Wind · Solar · Community development

In theory, the hallmarks of a successfully designed and implemented renewable energy policy are those that maximize RE grid-connected

© The Author(s) 2020
99
A. Lawrence, *South Africa's Energy Transition*, Progressive Energy
Policy, https://doi.org/10.1007/978-3-030-18903-7_5

generation at the intersection of the RE supply curve and the avoided cost of thermal electricity generation.[1] This formulation, however, begs the questions of how RE supply is determined, over what length of time and by what means it should be funded, how avoided thermal costs are assessed, and who decides. The policy vehicle that constitutes South Africa's first and only sustained attempt to harness its renewable energy endowment is its Renewable Energy Independent Power Producers Procurement Programme (REIPPPP). Prior added RE capacity does include the Sere Wind Farm and the country's first CSP plant, but at 100 MW each, these harnessed only a small fraction of the country's RE potential. Moreover, their financing—through the World Bank, African Development Bank, and Clean Development Fund—is, atypically, principally from public lenders.

Indeed, apart from these two installations, limited development of hydroelectricity on the Orange River, and nuclear power in suburban Cape Town (discussed in the previous chapter), the REIPPPP remains the only attempt to diversify sources of electricity generation. The Programme arguably represents a rare instance of successful policy coordination among across different branches of government. The REIPPPP benefited from a more favourable policy space, set of institutional arrangements, and operational factors that combined to create sufficient policy certainty on the role of renewable energy, as a green light to the private sector's investment strategy. Its realization ensued from coordination among the DoE, National Treasury, departments of Trade and Industry and of Environmental Affairs, Nersa, and Eskom, as well as financial institutions and project developers. In this regard, it has enjoyed advantages that the abortive FIT attempt (discussed in Chapter 3) clearly did not.

This level of coordination is unusual for the post-apartheid era, and took many years to coalesce. The 1998 Energy Policy White Paper envisioned a 70:30 split between Eskom and independent private-sector producers, introducing the principle of IPP participation for the first time. The 2003 White Paper on the Renewable Energy Policy of the Republic of South Africa set its objective of generating 10 megawatt-hours (MWh) of renewable energy by 2013 (approximately 4% of the energy mix); although initially unclear as to whether this referred to a cumulative or annual target and what sources would be included, the Department of Energy subsequently stated the target would be met by a combination of biomass, wind, solar,

---

[1] Meier et al. (2015: xviii).

and small hydroelectricity. As discussed in Chapter 3, however, the following year, the government abandoned plans for a competitive wholesale market in favour of retaining Eskom's exclusive single-buyer role.

The Integrated Resource Plan for Electricity 2010–2030 aimed to supply 42% of the new additional capacity over the 2010–2030 period, or 9% of the total generated electricity by 2030, from RE producers.[2] After inaugurating a compulsory bidders' conference with over 1000 participants bidding to provide 3625 MW of new capacity, the Programme procured its first MW of generation capacity in 2011, and by 2018 generated over 15MWh, including 7.7 MWh from wind. In terms of attracting FDI, the REIPPPP has been successful. After a late start, South Africa joined the top ten countries globally in terms of amount spent on RE investments. From 2011–2018, the REIPPPP has awarded 102 projects to the private sector, financed by R141 billion from domestic sources and R53 billion FDI, with a projected generating capacity of 6376 megawatt peak (MWp). The scheme has also helped bring generating prices down. Over the first three bidding phases lasting 30 months, average solar photovoltaic (PV) tariffs decreased by 68% in nominal terms and those for wind by 42%.[3] Over the subsequent two bidding phases lasting 27 months, PV tariffs decreased a further 47%, and those for wind, a further 29%.

Nonetheless, the REIPPPP has failed to make substantial inroads in mitigating, if not reversing, not only the high emissions, but also (as detailed in Chapter 2) the vicious cycle of poverty, inequality, and unemployment driven by the mining-centred economy. As this chapter shows, its design served to supplement, rather than replace, coal generation. Similarly, the goals of promoting industrial development, increased employment, and the constitutional right to clean and affordable necessities such as energy access are subordinated to that of creating an investment vehicle attractive to local and transnational financial interests.

Unlike Feed-in-Tariffs (FITs), tendering or bidding systems allow the government to control the amount of RE generation. As shown in Chapter 3, this control means that Eskom and other sites of policymaking are prone to insider-deals and influence peddling. Competitive bidding is intended to incentivize prospective providers to cut up-front costs, passing savings onto consumers and reducing investor profit incentives, but

[2]DoE (2011).
[3]Eberhardet al. (2014: 1).

also maintaining the consumer-producer divide. Yet despite these significant tariff reductions, the *retail* price of electricity for average consumers has outstripped inflation, growing 440% in real terms over the past decade, from under 25c/kWh to around 140c/kWh. While the increased tariffs are meant to cross-subsidize the introduction of RE sources, given the opacity of Eskom's governance, it is entirely possible that they subsidize coal generation, inflated salaries, and insider-deals to an even greater extent.

Indeed, in 2014, Eskom has used the same bidding system framework to introduce its so-called "baseload coal" IPP programme, with plant size capped at 600 MW. The winning bidders under the first bid window, Thabametsi near Lephalale, Limpopo (557 MW) and Khanyisa near eMalahleni, Mpumalanga (306 MW) are meant to provide short-term capacity increases. Yet in addition to being more expensive and, of course, far more polluting, they are unlikely to provide the intended capacity in less than the time that new, similarly sized RE sources have proven capable of being added to the grid; nor is it even clear that such increases (from any source) are necessary in the short term.

A report by the University of Cape Town's Energy Research Centre (ERC) shows that the additional incurred cost is almost R20 billion, which would crowd out cheaper and cleaner alternatives. By increasing GHG emissions by 205.7Mt CO_2e over the 30-year period of the power purchase agreements, the new plants would virtually wipe out the government's emission mitigation plans, including its expected savings from the entire Energy Efficiency Strategy to 2050. Since they have higher latent capacities already granted environmental authorisation (Thabametsi 1200 MW, and Khanyisa 600 MW), at these levels, the entailed costs and impacts would nearly double.[4] As Bobby Peak, Director of environmental justice NGO GroundWork observes, "Eskom is facing a financial crisis and rising electricity prices will drive consumers away from the utility. But the enormous additional expense of the coal IPPs would impact most directly on the poor, who are already hardest hit by the devastating health and other impacts of polluting coal-fired power generation." Although required to commence operating by December 2021, neither project has reached financial close yet, and because numerous required licences and authorisations are outstanding or currently being challenged in the High Court, neither is likely

---

[4] Ireland and Burton (2018).

to meet its operational deadline, particularly after Standard Bank decided against providing the necessary financing.[5]

An additional and related problem is that Eskom faces little incentive to treat IPPs fairly. From 2015 up to 2018, Eskom displayed a strong reluctance to issue budget quotes, which are a prerequisite for IPPs' bids to reach financial close, for the connection of new renewable energy IPPs to the grid—perhaps as a negotiation strategy to slow the development of IPPs and obtain additional funds from the regulator.[6] In May 2016, Eskom's then-CEO Brian Molefe publicly indicated his unwillingness to sign further PPAs with REIPPs. He complained (without providing evidence) that Eskom was being forced to sign twenty-year power purchase agreements at a higher cost than coal for technologies that would become obsolete by the time they were transferred to Eskom.[7] By that time, CSIR calculated that saved diesel costs due to RE sources alone totalled R7.2 billion; as we have seen, total RE costs have continued to decline since then. Regardless, Molefe's refusal meant that RE investment under the REIPPPP went from R170 billion in 2011 to R50 million in 2017. Only after Molefe had been replaced by Phakamani Hadebe in 2018 did the government substantially resume signing new RE contracts worth R56 billion, for 2300 MW over five years.[8]

There is little evidence, however, that the REIPPPP has thus far succeeded in significantly reducing unemployment levels. This is both because of the tacit cap on RE sources in government energy planning up to 2018, and because of the highly bureaucratic, competitive and secretive nature of the procurement process. Intended to incentivize bidders to promote job growth, domestic industrialization, community development, and Black Economic Empowerment (BEE) goals, the Programme's economic development requirements account for 30% of total bid value. This is a higher percentage than the Preferential Procurement Policy Framework Act (PPPFA)'s 10% implemented in 2000; it also emphasizes job creation (where it can be expected to have some impact, given the rural sites of most utility-scale projects) rather than the focus of ownership characteristic of BEE policies.

---

[5] Energy News (2018).

[6] Montmasson-Clair and das Nair (2017: 195).

[7] Creamer (2016).

[8] Khumalo (2018).

Prospective bidders have regularly complained that these stipulations are unduly burdensome and fraught with bureaucratic complexity. On the other hand, the "thresholds" at which bids are deemed compliant are generously low. Predictably, the actual bids in the first three rounds most often barely clear these thresholds (and in the case of concentrated solar power, fell short), rather than approach the higher aspirational "targets" (see Table 5.1).

More importantly, there is little oversight or accountability in terms of how these criteria are met. According to DoE official statistics, the following is the provisional result for the main RE sources in terms of local content and employment creation (Table 5.2).

Even under the assumption that these job totals are non-overlapping, the three main RE tender sources have generated slightly more than 54,000 jobs, of which about 20,000 are temporary construction jobs: far short of the government's goal from the 2011 COP17 meeting in Durban of creating 300,000 new jobs in South Africa's "green economy" by 2020, let alone the New Growth Path strategy document's goal of 5 million jobs by 2020.[9] But even this assumption may be overly optimistic, since the DoE sometimes equates the categories of "Person-months (PMs)", "Person-years (PYs)", or "full-time equivalent" (fte) with jobs themselves, making it difficult to extrapolate actual jobs from these subcategories.

Moreover, because the bidding tenders are targeted at utility-scale bidders rather than smaller scale ones, there is an in-built bias in favour of generating capacity over job creation, since smaller projects tend to create more jobs per MW of energy generated. As one local study found, large projects generate 5.83 jobs per MW on average, medium-scales ones, 5.3–8.0 jobs per MW (or 6.7 on average), and small-scale ones, 6.1–9.2 jobs per MW (or 7.7 on average).[10] Smaller projects, *ceteris paribus*, and easier to tailor to local conditions and political cultures, and more amenable to exercising greater levels of community oversight and control. At a time of growing slack in generating capacity and massive unemployment, the political case for the most employment-intensive approach to energy policy should not be difficult to make.

---

[9] Since aggregate employment has increased by fewer than 2 million since 2011, more than 3 million new jobs would need to be created over the next year for this goal to be met.

[10] EScience Associates et al. (2013).

Table 5.1 Elements of the REIPPPP economic development criteria[a]

Element (weighting)	Description	Threshold (%)	Target (%)
Job creation (25%)	RSA-based employees who are citizens	50	80
	RSA-based employees who are black people	30	50
	Skilled employees who are black people	18	30
	RSA-based employees who are citizens and from local communities	12	20
Local content (25%)	Value of local content spending	40–45[b]	65
Ownership (15%)	Shareholding by black ownership in the seller	12	30
	Shareholding by local communities in the seller	2.5	5
	Shareholding by black ownership in the construction contractor	8	20
	Shareholding by black people in the operations contractor	8	20
Management control (5%)	Black people in top management	–	40
Preferential procurement (10%)	BBBEE procurement[c]	–	60
	QME & SME procurement[c]	–	10
	Women-owned vendor procurement[c]	–	10
Enterprise development (5%)	Enterprise development contributions[d]	–	0.6
	Adjusted enterprise development contributions[d]	–	0.6
Socio-economic development (15%)	Socio-economic development contributions[d]	1	1.5
	Adjusted socio-economic development contributions[d]	1	1.5

[a]DoE (2014). *Source* Eberhard and Naude (2016), Table 3
[b]Depending on technology: 45% for Solar PV, 40% for all other technologies
[c]As percentage of total procurement spend
[d]As a percentage of revenue

Table 5.2 REIPPPP Economic Development Outcomes[a]

Technology	Round 1	Round 2	Round 3	Total
Solar PV				
Local content %	38.4	53.4	53.8	
Local construction jobs	2381	2270	2119	6770
Local operations jobs	6117	3809	7513	17,439
Wind energy				
Local content %	27.4	48.1	46.9	
Local construction jobs	1810	1787	2612	6209
Local operations jobs	2461	2238	8506	13,205
Concentrated solar power				
Local content %	34.6	43.8	44.3	
Local construction jobs	1883	1164	3082	6129
Local operations jobs	1382	1180	1730	4292
Employment grand total				54,044

[a]Adapted from Eberhardet al. (2014), Table 5

The rigor required to meet evaluation criteria for each step in the bidding process has served both to legitimate the Programme as well as mire it in technicalities, and has proven to be extremely time consuming and expensive. Paying key advisors such as legal experts can eat up fifteen percent of project development costs.[11] Cost reduction mechanisms to prevent the winner's curse phenomenon (wherein the winner will tend to overpay due to incomplete information) have constituted a hindering factor for the participation of new and/or smaller, previously disadvantaged players. The setting of ambitiously low price caps—although thereby reducing the complexity of price setting for the government and allowing prices to decrease rapidly as a response to increased competition—has deterred effective participation for some technologies, such as landfill gas and biomass (detailed further in the next chapter), where a smaller scale would have improved the chances of community participation.[12]

Effective community participation is deterred in other respects as well. To encourage social development in sites of RE projects while deterring nepotism, community trusts are mandated, made up of members living

[11] Montmasson-Clair, Moilwa and Ryan (2014).

[12] Montmasson-Clair and das Nair (2017).

within a 50-km radius of the project site.[13] As funded by the Development Bank of Southern Africa (DBSA), the Industrial Development Corporation (IDC)and/or the Public Investment Corporation to the amount of R9.5 billion collectively over the first three bid windows, these trusts bestow communities with a stake of up to 5% on average, per project. These often entail low-interest financing to community trusts to buy shares in the project company that are managed by the DBSA and the community trust leadership, who jointly decide on how the revenue is to be spent.[14] Most of the promised income for communities thereby generated, however, will not even become available until after 2028.[15]

Furthermore, without having clear developmental objectives—partly because of their bypassing prior social trust networks—these community trusts' management practices have created some unintended consequences. Concerns have been raised about the concentration of these funds in a limited number of communities, their monitoring and evaluation, and the capacity of the DoE and development finance institutions to manage the funds and ensure IPPs meet their commitments. Moreover, many community trusts are viewed as have been established merely to serve the requirements of the request for proposals: these trusts engender mistrust. Project developers and the local development finance institutions have little experience in working with communities and municipalities in these areas to ensure that development programmes are aligned with community interests and municipal plans.[16]

The risk is that such community trusts will receive excessive financial flows with little knowledge of the communities in which they are working. An implementation agreement signed with the DoE is meant to ensure that preferred bidders adhere to their commitments. Each bidder is required to report to the DoE on a quarterly basis regarding these commitments.[17] The REIPPPP awards more points to communities located closer to the renewable energy project and does not place a limit on multiple community trusts for one community. This results in a small number of communities in lucrative areas for RE being assigned multiple community trusts, thereby

---

[13]Van den Berg (2013).

[14]Montmasson-Clair, Moilwa and Ryan (2014).

[15]McDaid (2016: 16).

[16]Montmasson-Clair and das Nair (2017: 208–210).

[17]DoE (2012).

subordinating development to project-driven siting rationales.[18] In other cases, the community-owned dividend only earns income after the debt on the project is paid off, sometimes after a period of fifteen years or more.[19]

The IPP process envisioned no formal role for local metro and municipality government representatives, with whom it engages at best in a fragmented, peripheral and ad hoc manner. As elected representatives of the communities in question, local government officials are the obvious authorities to monitor IPPs' conformity to developments targets.[20] Together with the lack of transparency surrounding the selection of socioeconomic projects, representatives' exclusion exacerbates their mistrust with councillors; in some cases, developers even attempted to retain control of the trusts, under the guise of "ownership" by the community.[21] The appointment of the trustees and the management of the funds remain problematic and appear far from the standards of good governance, with evidence of nepotism, political arrangement, elite capture and lack of communication, transparency and accountability.[22] Meanwhile, community members reasonably expect the delivery of more immediate benefits, not least, improved access to electricity, which (as shown in Chapter 2) is more often inadequate in the rural areas where REIPPPP projects are typically located. As one government official remarked, "communities want electricity the same as in urban areas and if they don't get it they will get very frustrated. We underestimate communities but they are lot more aware than we think and sometimes more aware than we are. So the biggest thing besides with the technical stuff that could go wrong… is with community interaction."[23]

Many such problems stem from a disinclination on the part of RE companies to take in-depth, extended engagement with communities sufficiently seriously. As with mining, there is ample opportunity to do so, given the long-term nature of RE investment in a given project, typically last at least two decades, but such opportunities are typically not pursued. One study of a solar installation by Scatec Solar in Northern Cape province found, for example, that instead of engaging dialogically with the local

---

[18] Montmasson-Clair and das Nair (2017: 210–211).

[19] Baker (2015: 153).

[20] McDaid (2016: 38).

[21] McDaid (2014) and Tait et al. (2013).

[22] McDaid (2014).

[23] Quoted in Baker (2015: 154).

community using continuous, holistic and open-ended monitoring and evaluation, the RE vendor chose a tokenistic approach of "engaging" and "informing" community representatives. It implicitly equated the Community Development Trust with community legitimacy, neglected to assess local energy poverty levels and jointly devise means of addressing them, or even communicate sufficiently in the local languages and take into account the levels of local illiteracy.[24]

In addition, skilled positions are most often filled by personnel outside of the local community in question, and mechanisms of job allocation are opaque.[25] More fundamentally, the category of "job" serves as an ideological shell in this context, without significant attention paid to quantitative questions—such as their actual duration and comparability from one project to another—and at least as important qualitative ones, such as how meaningful the employment is for the worker in question and how valuable for the broader community. For all the complexity of the tendering process, this crucial dimension of information is not being adequately collected.

The problems associated with promoting employment growth and localized RE manufacturing that are intrinsic to the competitive bid-round structure will likely grow only more intense. Although the local content threshold kept increasing with each of the prior bidding rounds, with the hiatus imposed by Molefe, most local manufacturing plants did not have sufficient orders to survive. "There were several local manufacturing plants as of a couple of years ago. Now most of those have closed due to lack of demand," observed Chris Ahlfeldt, Cape Town-based energy consultant. This suggested that holding independent power producers to account for local content thresholds that were viable previously "may be difficult for Round 4" and could get progressively more so for Round 5, possibly requiring the government to commit to further bids in the future before manufacturers (particularly for PV plants, which require a relatively high level of precision-manufactured components) would feel confident in investing in facilities again.[26]

Finally, there is the question of whether the continued financing of the REIPPPP is sustainable on its current trajectory. As better-resourced foreign transnationals play an ever-larger role in securing contracts with each

[24]Shaw (2017).
[25]McDaid (2014).
[26]Deign (2018).

subsequent bidding round, smaller national players are increasingly priced out of the market. Using the vehicle of project finance, foreign RE firms are better able to control levels of risk and extract more favourable terms of return on their investments. By providing finance-based debt on fixed loan terms, lenders are not liable for any losses a given RE project may incur. At the same time, local sources of finances, such as South Africa's major banks in conjunction with its IDC, the DBSA, and to a lesser extent, international DFIs such as the World Bank's International Finance Corporation (IFC) and International Bank for Reconstruction and Development (IBRD), as well as the European Investment Bank (EIB), serve to mitigate currency-denominated risk that international banks generally shun, given the major fluctuations the rand experienced after abandoning currency controls. The higher cost of REIPPPP financing is reflected in the fact that interest rates tend to be nearly twice as high as those for similar EU or European projects.[27]

Whereas the initial bidding rounds were more profitable to investors than subsequent ones have been, given their increasingly competitive nature, the predominant debt financing from South African banks of the first two rounds has given way to a growing presence of corporate finance, which is subject to less stringent loan requirements. The trends raise several questions about the financial as well as political sustainability of the REIPPPP. Differential rules regarding the on-selling of debt v. equity and "empowerment" (BEE or community shareholder) v. non-empowerment shareholders tend to favour larger, transnational owners rather than smaller, local ones. Whereas debt can be on-sold immediately subject to DoE approval, equity can only be sold after three years, requiring both DoE and lender approval. Large private equity interests are able to de-risk the assets and on-sell at a profit to longer-term capital managers such as pension funds. To meet the economic development requirements of REIPPPP, projects must meet minimum thresholds of a 12% for BEE shareholding and 2.5% for community shareholding; owners of these share categories are therefore constrained from on-selling except to buyers with the same BEE score.[28] As we have seen in Chapter 3 in the case of Eskom's coal markets, there are very few companies with the technical know-how to conduct such trades and tend to strongly overlap with comprador elites holding strong

---

[27] Baker (2015: 150–152).

[28] Baker (2015: 154–155).

political connections. In bull markets, they are likely to benefit from these ties and reap disproportionate benefits compared with community members and the less well-connected. In bear markets, the larger transnational financial interests will be able to dump these assets far more quickly than local owners will be able to.

CONCLUSION

Chapter 3 detailed ways in which BEE legislation may have had an unintended, yet salutary, effect of slowing down the rate of coal mining and mining more generally, through its imposition of additional layers of red tape as well as insider dealing "tenderpreneurship". The same effect, however, applies to the country's embrace of renewable energy, since BEE strictures have tended to dampen investor enthusiasm and participation. Eskom's treatment of the programme has veered between erratic at best and prejudicial at worst, imposing further limits on the addition of RE capacity. Independently from these factors, however, the same is arguably true for the government-controlled competitive bidding structure of the REIPPPP. Each bidding round effectively reduces the margin of profit investors can achieve. The landscape of viable RE sites becomes progressively fragmented, reducing the capacity of the state to promote and plan for nationally or transnationally articulated RE infrastructure. The job creation track record is underwhelming. Worse, the uncertainty imposed by this structure has debilitated the capacity for local manufacturers of RE inputs to engage in longer-term planning, while the complexity of instruments used to finance the REIPPPP has rendered it increasingly vulnerable to exogenous financial shock and progressive loss of local control. Meanwhile, coal interests have used the programme as a vehicle to further increase unnecessary, expensive, and damaging coal generation for private gain.

The combination of local and national political as well as fiscal sources of unsustainability and muted environmental benefit does not augur well for the REIPPPP as a whole. It is for this reason that the Alternative Information Development Centre (AIDC) recommends that the REIPPPP be terminated in its current form after Round 5 and replaced with a plan aiming to attain 100% RE generation in the shortest possible period, return socio-economic development competency to the government (and to empowered local authorities in particular), with much more detailed and transparent monitoring of REIPPPP during the interim.[29]

In a sense, the logic of the REIPPPP structure is fundamentally similar to the one guiding Eskom's 1987 restructuring and the 2002 Eskom Conversion Act, whereby the non-for-profit mission of the parastatal was replaced with one of income generation. This for-profit orientation explains the principal objection that the metalworkers' union NUMSA gave in their May 2018 court challenge to prevent the Minister of Energy from signing 27 REIPP contracts.[30] Although unsuccessful in this immediate goal, the challenge partly served to critique the discourse of energy as a commodity, arguing instead that its provision should be seen as a basic human right. As NUMSA Deputy General Secretary Karl Kloete explained at the time, a resolution on the Socially Owned Renewable Energy Sector in South Africa adopted at NUMSA's 2012 National Congress stated that

> the country's energy needs should be met by a mix of different forms of collective ownership including energy parastatals, cooperatives, municipal-owned entities and other forms of community energy enterprises …[involving] some level of decentralised ownership and operation integrated into a coherent, national centre. We made it clear that the national grid must be publicly-owned and must remain the backbone of energy provision [and] that the mandate of Renewable Energy projects must be to achieve service provision, meet universal needs, decommodify energy and provide an equitable dividend to communities and workers directly involved in production and consumption of energy. We stressed that socially-owned RE enterprises should be non-profit entities. … that the producers and owners of the means of renewable energy production must receive a large share of the sector's economic benefits, [and that] renewable energy has great potential to give communities greater control of their resources and to satisfy their energy needs on a decentralized basis. We were, and remain, absolutely clear that renewable energy is essential to mitigate climate change and that the renewable energy sector is not developing as fast as it needs to and that a socially-owned and controlled sector can push its development far faster than it is currently expanding.[31]

---

[29]McDaid (2016: 36–37).

[30]NUMSA is the country's largest single union and largest affiliate of the South African Federation of Trade Unions (SAFTU), the confederation that broke away from Cosatu, the South African Communist Party, and the ANC in 2017.

[31]Cloete (2018).

The concluding chapter critically examines options for such a collectively owned, decentralized and not-for-profit renewable energy infrastructure as a means of simultaneously addressing South Africa's ecological, economic, unemployment, and political legitimacy crises.

REFERENCES

Baker, L. (2015, August). The evolving role of finance in South Africa's renewable energy sector. *Geoforum*. https://doi.org/10.1016/j.geoforum.2015.06.017.

Cloete, K. (2018, March 15). Our problem with the IPPs. *PoliticsWeb*. Accessible at: https://www.politicsweb.co.za/opinion/our-problem-with-the-ipps.

Creamer, T. (2016, July 21). Eskom letter sends shock waves through private power sector. *Creamer Media's Engineering News*. Accessible at: http://www.engineeringnews.co.za/article/eskom-letter-sends-shock-waves-through-private-power-sector-2016-07-21.

Deign, J. (2018, July). *South Africa Open Again for Renewables After Auction Turmoil, Greentech Media*. Accessible at: https://www.greentechmedia.com/articles/read/south-africa-open-again-for-renewables-after-auction-turmoil.

Department of Energy (DoE). (2011). *South African Department of Energy, Integrated Resource Plan for Electricity, 2010–2030*.

Department of Energy (DoE). (2012, September 27). Postponement of 3rd bid submission date for the REIPPP. *EE Publishers*. Accessible at: http://www.ee.co.za/article/deptenergy-359-10-postponement-of-3rd-bidsubmission-date-for-the-reipppp.html.

DoE (Department of Energy). (2014). *The REIPP Procurement Programme Part B: Qualification criteria*. Republic of South Africa.

Eberhard, A., Kolker, J., & Leigland, J. (2014). *South Africa's renewable energy IPP procurement program: Success factors and lessons*. Accessible at: https://www.gsb.uct.ac.za/files/ppiafreport.pdf.

Eberhard, A., & Naude, R. (2016). The South African renewable energy independent power producer procurement programme: A review and lessons learned. *Journal of Energy in Southern Africa, 27*(4). Accessible at: http://www.scielo.org.za/scielo.php?script=sci_arttext&pid=S1021-447X2016000400001.

Energy News. (2018, May 30). Report: SA doesn't need R19bn coal IPP projects. *Energy News*. Accessible at: https://www.bizcommunity.com/Article/196/693/177655.html.

EScience Associates, UrbanECON, & Ahlfeldt, C. (2013). *The localisation potential of photovoltaics (PV) and a strategy to support large scale roll-out in South Africa* (Report prepared for SAPVIA, DTi, WWF, South Africa). Accessible at: http://www.sapvia.co.za/wp-content/uploads/2013/04/PV-Localisation_Draft-FinalReport-v1.2.pdf.

Ireland, G., & Burton, J. (2018). *An assessment of new coal plants in South Africa's electricity future: The cost, emissions, and supply security implications of the coal IPP programme.* Energy Research Centre, University of Cape Town, Cape Town, South Africa. Accessible at: https://cer.org.za/wp-content/uploads/2018/05/ERC-Coal-IPP-Study-Report-Finalv2-290518.pdf.

Khumalo, S. (2018, April 4). Jeff Radebe signs R56bn contract with renewable power producers. *Fin24.* Accessible at: https://www.fin24.com/Economy/Eskom/jeff-radebe-signs-long-delayed-renewable-power-deals-20180404.

Meier, P., Vagliasindi, M., & Mudassar, I. (2015). *The design and sustainability of renewable energy incentives: An economic analysis* (pp. 155–169). Washington, DC: World Bank Group.

McDaid, L. (2014). *Renewable Energy Independent Power Producer Procurement Programme Review 2014.* Cape Town: Electricity Governance Initiative of South Africa.

McDaid, L. (2016). *Renewable energy independent power producer procurement programme review 2016: A critique of process of implementation of socio-economic benefits including job creation.* Cape Town: AIDC.

Montmasson-Clair, G., & Das Nair, R. (2017). South Africa's renewable energy experience: Inclusive growth lessons. In J. Klaaren, S. Roberts, & I. Valodia (Eds.), *Competition Law and Economic Regulation in Southern Africa: Addressing Market Power in Southern Africa* (pp. 97–119). Johannesburg: Wits University Press.

Montmasson-Clair, G., Moilwa, K., & Ryan, G. (2014). *Regulatory Entities Capacity Building Project Review of Regulators Orientation and Performance: Review of Renewable Energy Regulation.* Johannesburg and Pretoria: University of Johannesburg and Trade and Industrial Policy Strategies.

Shaw, J. (2017, March). *Assessing the sustainability of an independent power producer's social investment in a community: A case study of Scatec Solar* (MPA dissertation). Stellenbosch University, Stellenbosch.

Tait, L., Wlokas, H. L., & Garside, B. (2013). *Making communities count: Maximising local benefit potential in South Africa's Renewable Energy Independent Power Producer Procurement Programme (RE IPPPP).* Cape Town: International Institute for Environment and Development.

Van den Berg, J. (2013). Submission to NERSA: Eskom MYPD 3 Application. *South African Renewable Energy Council.* Accessible at: http://www.nersa.org.za/Admin/Document/Editor/file/Consultations/Electricity/Presentations/South%20African%20Renewable%20Energy%20Council.pdf.

CHAPTER 6

Conclusion: Just an Energy Transition—Or a Just Transition?

Abstract A just and sustainable energy transition in South Africa would entail not only the active mobilization of affected community members in developing energy and infrastructure at scales, in communities, and from sources that have been thus far neglected. It would also explicitly account for the immense and various sources of waste, fraud, abuse and degradation that are intrinsic to the country's extractive status quo. A hint of what active community mobilization around energy justice can achieve is provided by Operation Khanyisa; but it is unclear whether this success can be broadened nationally or sustained further. The Million Climate Jobs (MCJ) campaign provides a realistic alternative approach to transforming the country's energy system and broader political economy. Two additional, and potentially complementary, policy options are discussed: state guaranteed employment in the climate jobs sector, and a progressive feed in tariff (FIT).

Keywords Just and sustainable energy transitions · Operation Khanyisa · Million Climate Jobs campaign · Feed in tariff (FIT)

On 12 March 2018, five-year-old Lumka Mkhethwa went missing from Luna Primary School in the Eastern Cape and was found the next day

© The Author(s) 2020
A. Lawrence, *South Africa's Energy Transition*, Progressive Energy
Policy, https://doi.org/10.1007/978-3-030-18903-7_6

drowned in the school's two meter-deep pit latrine, its only toilet facility.[1] This was not an isolated case. In January 2014, during his first week at school, five-year-old Michael Komape drowned in a pit latrine in Limpopo province. Four years later, in July 2018, not far from the site of Komape's death, three-year-old Omari Monono suffered the same fate.[2] In the wake of the R1.5 trillion lost in headline-grabbing "state capture" corruption, and far less-visible billions lost to licit and illicit financial outflows, the tragic symbolism of these children and too many others of their generation drowning in a cesspool of corruption and neglect is inescapable.[3]

Given that more than a third of all schools nationwide—more than 9000 out of nearly 25,000—have pit latrines, including about 4000 that lack any other sanitary facilities, these deaths were all but inevitable. Together with the hundreds of schools lacking water, electricity, or both, they are concentrated in the poorest rural areas of KwaZulu-Natal, Limpopo, and the Eastern Cape. Indeed, in the Eastern Cape alone, 61 schools still lack any toilets at all, a deprivation shared by nearly 15 million South Africans. Whereas the government categorizes pit latrines as "basic sanitation," the charity WaterAid estimates that 27% of the population lack even this level of provision.[4] In 2013, the Department of Education had set November 2016 as the deadline by which it would oversee the complete replacement of all unsafe and inappropriately built schools; two years later, an immense backlog remains. Basic Education Minister Angie Motshekga cited budget cuts as among the reasons why improvement of school infrastructure has been slow; a further R3.5 billion is expected to be cut from the infrastructure initiative in the next three years.[5]

This deficit is part of the national crisis in service delivery, including sanitation. Pit latrines make a mockery of the right to proper sanitation, part of a basic standard of living enshrined as a human right in South Africa's constitution. Mkhethwa's death prompted President Cyril Ramaphosa to call for the eradication of pit latrines in all schools, giving the education department three months to come up with a plan. An interim government report calculated that converting all school pit latrines into flush toilets will

---

[1] BBC (2018a).

[2] BBC (2018b).

[3] Merten (2019).

[4] WaterAid (2017: 26).

[5] Saba et al. (2018).

cost more than R6.8 billion, or about R700,000 per school. This pattern of excessive cost of services delivered at a snail's pace is also found in Eskom's housing electrification programme, which manages electrification of only about 160,000 homes per year on a budget for 2015–2020 of R10 billion. The crisis of service delivery, in turn, is implicated in the broader crisis of democratic governance, a morbid symptom of the "national democratic revolution" having ground to a halt while viable alternatives are not yet fully born.

For the designation of "just and sustainable transitions" to be meaningful, it must prioritize schools and communities like Lumka's, Michael's, and Omari's, and provide practical alternatives—for alternatives certainly exist. A month after Lumka's death, a school whose electricity had been disconnected by Eskom, 486 pupil-Poelano Secondary School in Northwest Province, went off-grid with a solar PV and hydrogen fuel cell (HFC)-powered system that meets the school's lighting and IT needs. Developed in partnership with the Centre for Scientific and Industrial Research (CSIR), North-West University, University of the Western Cape, University of Cape Town, and Mintek, this inaugural system cost just under R10 million (including the solar PV panels, fuel cell system, electrolyser for on-site hydrogen production from water, hydrogen storage system and battery system for electrical power storage). But since the total cost includes large up-front R&D costs, the replicated models planned for more than 5000 off-grid rural schools, clinics, and homes are expected to benefit from economies of scale—perhaps costing less than the R700,000 per school the government anticipates spending for pit latrine conversion.[6]

A yet lower-cost option would be to connect the schools (as well as clinics and villages) to biodigester infrastructure, as EarthLife Africa advocates, which yields the additional benefits of providing methane for heating, refrigeration, and cooking without smoke, and fertilizer in the form of bioslurry (thereby helping reduce costs and damage from chemical fertilizers), as well as improved soil and sanitation. The equipment is readily affordable with microloan assistance and pays itself back in gas and fertilizer outputs alone, typically in under five years. It is suited to urban as well as rural areas: Earthlife recently supervised the installation of a digester at Khangezile Primary School in KwaThema, eastern Gauteng province, as well as three other pilot schools. Together with rooftop solar PV, it enables

---

[6]Alfreds (2018).

the school to go virtually off-grid, generating gas for cooking two free school meals a day for all 400 pupils.[7] The potential for scaling up and promoting the mutual gains of reducing malnutrition while strengthening energy self-sufficiency and small farmers' livelihoods is clear.

According to a recent estimate, biogas could easily generate 2.5 GW of electricity (more than is provided for by South African nuclear power, or by all the REIPPPP projects to date put together). It could provide much of the peaking power needed to complement RE sources in the short term, with an internal market potential of R10 billion that could create thousands of jobs with high levels of local ownership and control. Already, larger existing plants include the 1.2 MW Johannesburg Northern Waste Water Treatment Works biogas plant and the 4.4 MW Bronkhorstspruit plant, Africa's largest biogas plant. South Africa's cattle industry alone could generate 1.26 tWh of electricity by this means per year.[8] To date, however, there is virtually no market penetration, with only 400 units of all sizes (half of which were installed in the 1970s and 1980s to improve the running of sewage treatment plants). This is fewer than in neighbouring Lesotho, with less than 5% of South Africa's population and under 1% of its GDP. As we have seen in Chapter 5, the REIPPPP is structurally biased against the scale of such smaller generating projects, irrespective of their potentially lower costs and greater benefits. The barriers have less to do with feasibility or cost than with political leadership and awareness of alternatives.

This basic assessment pertains not just to the removal of deadly hazards from and electrification of schools, clinics, and housing in rural areas, but extends to South Africa's stalled energy justice transition as a whole. The primary if not sole constraint is not natural, fiscal, or technical, but political. Not only does South Africa's energy need to be politicized, but its politics need to be energized. The question of affordability, moreover, needs to be placed in a broader context, one that critically scrutinizes past spending proposals and outcomes and more fully accounts for opportunity costs.

An assessment of past spending should begin with a full accounting of the various direct and indirect subsidies that fossil fuel and nuclear energy

---

[7] BiogasSA sells similar biodigester units for R12,000–R42,000, and has expanded to industrial units of up to 0.05 MW. ENCA (2015).

[8] With over a million heads of cattle, South Africa's cattle industry alone could easily accommodate 579,000 plants, with an estimated total biogas rate of over 67,000 (m3/h), a potential calorific power of 3.6 tWh/yr and estimated electricity potential of 1.26 tWh/yr (Roopnarain and Adeleke 2017: 1174; Okudoh et al. 2014).

have entailed. Subsidies in the form of foregone revenue, price support, and transfers from 2008 to 2015 alone have ranged from $675 million to nearly $2 billion per year, or cumulatively, over $7 billion. Sasol alone likely accounts for over $100 million per year; it would not exist and could not survive without massive state and consumer subsidies, and its liquid fuels pricing regime ensures large profits for its coal-to-liquids business to the detriment of consumers. Although a full inventory taking the various dimensions of exploration, extraction, distribution and consumption of fossil fuels into account is difficult due to incomplete information, Eskom's loans, loan guarantees, transfers, water infrastructure subsidies and diesel rebates together from 2008 to 2015 alone have ranged from over $500 million to nearly $2 billion per year, or cumulatively, over $7 billion (roughly R100 billion).[9]

Together, these costs constitute a subsidy to a hazardous and unsustainable sector of over R200 billion during the Zuma presidency alone. It bears underscoring that these billions subsidized a service hobbled by a regular series of blackouts and load-shedding that have cost as much as R1.4 trillion over the past decade and are forecast to lose between a third and a half of GDP for the first quarter of 2019, or between R1.59 to R2.4 trillion.[10] To these costs should be added the fallout from state capture during the Zuma years, including the R200 billion wasted on Medupi and Kusile mega coal plants (discussed in Chapter 2), more than R400 billion in lost budget, debt service costs, and uncollected tax revenue, and more than R800 billion in wiped out share and bond value.[11]

The broader costs of not just coal mining, but a mining-dependent economy, however, are even greater. The negative health externalities of coal mining were touched on in Chapter 1, but their broader scope is readily inferred from the country's leading causes of death. These may be thought of as a series of concentric circles expanding outward, from the aggregate morbidity and mortality that has accompanied the extraction, processing, and distribution of ore; to the extended effects on social reproduction of the miners, their families, and communities; to the political, institutional, and cultural effects of mining on the country, the region, and beyond.

[9]Burton et al. (2018: 232–235).
[10]Head (2019).
[11]Merten (2019).

Although it is impossible to arrive at a precise figure for the cumulative figure for deaths from the extraction, processing, and distribution of ore, in gold mining alone, it numbers in the hundreds of thousands since the late nineteenth century, at a rate than has improved all too slowly over the past century.[12] The death toll from mining-related diseases is even higher: tuberculosis (TB) continues to be the leading cause of death in South Africa, conservatively estimated at 25,000 deaths a year (but much higher when its contributory role to the more than 100,000 AIDS deaths per year is accounted for). At any given moment, perhaps 80% of the population of South Africa is infected with TB bacteria. Nearly 1% of the country's population develops active TB each year (roughly 5% of the global active TB population); in most provinces, it accounts for close to 10% of all deaths (more, in KwaZulu Natal).[13] As with AIDS, it has long been implicated in the mining economy—where the TB incidence rate is almost three times that of the general population and ten times higher than the rate the WHO classifies as an emergency—and in the substandard quality of housing in the townships and squatter camps.[14]

In the words of a May 2016 South Gauteng High Court judgment in a class-action suit brought by miners' representatives on behalf of as many as 500,000 claimants against the major mining houses, South Africa's gold mining industry "left in its trail tens of thousands, if not hundreds of thousands, of current and former underground mineworkers who suffered from debilitating and incurable silicosis and pulmonary tuberculosis."[15] Despite recent government tenders bringing drug costs down, new TB drugs can cost R36,000 per patient to treat each year.[16] The newer drug-resistant TB strains, moreover, can cost South Africa over R200,000 per

---

[12] Its fatality rate improved by only 33% in the five decades after World War II (whereas the diamond mine fatality rate showed no improvement for seven decades from the 1920s to 1990s). Although South Africa's worst mining disaster was the 1960 Coalbrook colliery collapse, killing 437 miners, gold mining continues to have an even worse fatality rate. As late as the 1990s, according to one estimate, over a career working two decades underground, the average South African miner (of all ore types) faced a one in thirty chance of dying in an occupational accident. These risks are not merely a result of general mining hazards, but more specifically derive from the racialized violence of South Africa's industry.

[13] Kanabus (2018).

[14] See World Bank's (2017).

[15] South Gauteng High Court (2016).

[16] Gonzalez (2016).

patient to treat, and even so, most courses of treatment are not successful.[17] If all TB sufferers were to receive adequate treatment, the cost would be over R15 billion per year. Even so, treatment costs are less than half of total losses to the economy from TB, which include such indirect costs as lost income, costs to guardians and care-givers, and costs to the broader health system.[18]

As noted in Chapters 2 and 3, another major cost strongly associated with the Mining Energy Complex is capital flight. Even before the Guptas got in on the act, the annual level of capital flight approached 20% of GDP, or $70 billion (nearly R1 trillion). As long as MEC firms dominate mining and mining remains the dominant sector, the continued loss of hundreds of billions of rands each year remains a virtual certainty. The connections of this seepage to unemployment, inequality, poverty, and violence should be blindingly obvious.

The costs of such capital flight can be seen, for example, in the case of crime; as is widely recognized (if all too often overlooked), inequality is a leading cause of violent crime the world over.[19] South Africa has the fifth highest homicide rate in the world.[20] StatsSA ranks assault as the second leading cause of non-communicable death; the rate of femicide—the killing of women and girls—is five times the global average.[21] One estimate of the costs of crime alone was almost 8% of GDP, i.e. over $5 billion USD/R63 billion.[22] More recently, analysis from the Institute for Economics and Peace (IEP) found that the national cost of violence in South Africa is equivalent to 19% of the country's GDP—the sixteenth highest rate in the world (up from their 2015 report, where South Africa ranked 33rd), with total violence containment spending in South Africa amounting to $66.7 billion (R989 billion) in 2016 (about R1.84 trillion in PPP terms, roughly R34,160 per capita).[23] Taking into account health, welfare, and economic factors, then, the sum total from this incomplete list of opportunity costs includes tens of thousands of needless deaths and a trillion of stolen or

---

[17] Hello Doctor (2017).

[18] Foster et al. (2015).

[19] Fajnzylber et al. (2002: 1) and Krohn (1976).

[20] BBC (2018c).

[21] Makou (2018).

[22] Alda and Cuesta (2011).

[23] BusinessTech (2016).

misspent rands *each year.* The annual loss of life approaches or exceeds the total South African civilian and military deaths from the 1899–1902 South African war, World War I and World War II combined.

In light of these losses, a South Africa without fossil fuels as the dominant energy source—although necessarily entailing far-reaching transformations of South Africa's economy, institutions, and society—is an urgent imperative. Chapter 1 showed that despite official commitments endorsing clean energy policies, and despite an increasingly strong *prima facie* economic (no less than ecological) case for pursuing them, the South African government has thus far dragged its feet. Chapter 2 suggested that a deepened understanding of energy transitions as "just transitions" would account for transnational structures, as well as socio-economic and ecological imperatives. Chapter 3 illustrated the dynamics of the MEC through a focus on Eskom, arguing that the current discussions of (and investigations into) "state capture" should adopt a broader historical and institutional perspective—not to exonerate those guilty of political corruption, but to better understand the enabling environment in which they have operated. Chapter 4 further suggested that coal's competing non-renewable energy sources—fracking and offshore gas, as well as geothermal and nuclear power—are neither economically nor ecologically viable. At the same time, however, the REIPPP programme—although inaugurating large-scale renewable provision independent from Eskom—faces sustainability challenges of its own, as Chapter 5 details. The competitive bidding structure of the programme has entailed suboptimal outcomes in terms of promoting local manufacturing of inputs and thus in terms of employment creation more broadly, while those jobs that have been created are very unevenly distributed. In addition, contractual stipulations meant to empower local communities have often led to increased mistrust and conflict instead. Even worse, the government has used the IPP structure to inaugurate new coal generation contracts, rather than accelerate a RE transition. Most fundamentally, it has placed the attraction of foreign direct investment as an end to itself, rather than using RE generation as a means to realizing constitutional rights to a decent life for all.

It is entirely conceivable that the status quo—a vicious cycle of hastening capital flight, economic instability, fiscal austerity, massive unemployment, worsening poverty, inequality and ecological crisis—will continue in the short- to medium-term, however weakly mitigated by the modest pace at which the REIPPPP continues to be implemented. The energy system's overwhelming reliance on fossil fuels as the country's dominant energy source could be extended this way for years, even decades. This would amount to another passive revolution—or "just a transition".

A true alternative—a just transition to a transformed energy system, economy, and society—would strive to combine the most rapid feasible decarbonisation with the most rapid employment creation. At the same time, it would lay the groundwork for economic diversification—away from mining and extraction and toward strengthening needs-oriented manufacturing, food, and water systems as a means of strengthening South African communities' health and welfare. This, in essence, is the vision entailed by the Million Climate Jobs (MCJ) campaign, an alliance of trade unions, social movements and community organizations campaigning for policies that combine the climate justice imperative of reducing emissions with the energy and socio-economic justice imperative of improving livelihoods through employment creation. As discussed in Chapter 2, reducing unemployment can have a positive, direct, short-term effect on growth, of one percentage point or more GDP per year for every five percentage point reduction in unemployment.[24] Public works initiatives are arguably the best or only feasible means, and energy systems transformation, the best or only feasible end, to attain these goals simultaneously. Ultimately, such a program would be the most effective way finally to eradicate the legacies of apartheid that have concentrated social injustice and inequalities racially, spatially (in the both rural areas as well townships and informal settlements), and along gendered lines.

Not just the scale, but the type of state-guaranteed employment is of decisive importance. The Zuma presidency was a period of increased public sector employment of more than 1.5 million civil service positions, including a nearly 50% increase in total Eskom staff and more than 300% increase of staff salaries (averaging almost R700,000, or roughly three times the average formal non-agricultural sector salary). Yet these increases clearly have not decreased poverty or inequality, but on the contrary, contributed to a fall in the real rate of return on investment due to decreased productivity.[25] Other standard employment creation mechanisms do not in themselves direct employment creation toward sustainability goals. These

---

[24] See SAMI (2018).
[25] Klein (2012).

include reducing interest rates[26]; fiscal expansion[27]; increasing and broadening access to welfare benefits (such as increasing and transforming the current social grants into universal income grants); cutting payroll taxes for new hires[28]; or directly subsidizing employment. Nor do such measures (apart from universal income grant provision) address basic human needs.[29]

Climate jobs, by contrast, are those that combine environmental improvement and protection, increase energy and resource-use efficiency while deploying RE technology, and significantly reduce South Africa's poverty, inequality, and unemployment. Because they promote infrastructure spending and target the 40% of the working-age population that is unemployed, underemployed, or stuck in low-wage, precarious employment, these jobs would also have a greater multiplier effect than the government's prior efforts at employment creation.

Some of the main areas of climate jobs potential that would address South Africa's immediate needs along these lines are as follows (in rough order of total jobs created)[30]:

1. Transport (390,000)
2. Electricity and Renewable Energy (RE) (250,000)
3. Agriculture (100,000–500,000)
4. Construction/Eco-housing and Repairs (150,000–200,000)
5. Waste, industry, education/efficiency promotion (110,000)

These are conservative estimates; other estimates of job creation potential are substantially higher, including categories such as community caregivers (more than one million part-time positions), tourism, and

[26] Mosler (1997).

[27] Developing countries engaging in fiscal expansion have experienced a wide range of outcomes, so policy appears less linear than for OECD countries (see, e.g., Giavazzi et al. 2000).

[28] These could include Pay As You Earn (PAYE) and Standard Income Tax on Employees (SITE) levies to the South African Revenue Service (SARS), or gross revenue/salary-related levies to district councils. These may have particular benefits for poor South African women, albeit not as great as increased social grant expenditure; see Budlender et al. (2010). For an early comparative discussion, see Freeman (1992).

[29] On universal income grant provision in South Africa, see Marais (2018).

[30] Ashley (2018: 281).

ecological restoration (at least half a million each).[31] Each category combines large-scale employment creation with improved efficiency, sustainability, and pollution reduction, although deep decarbonization pathways and job creation are not proportional in each instance. For example, almost half the amount of the envisioned CO_2 equivalent reduction from these proposals, more than 230 million tons, comes from a coal-to-RE transition. The next biggest reduction, more than 70 million tons, comes from electrification of transport and increases in mass transit.[32] Nonetheless, in terms of health and human welfare improvements, each area has strong *prima facie* arguments in its favour. For example, under the fifth category, improving water access and treating acid mine water drainage have become areas of growing urgency; locally developed solutions are available, with the potential to create thousands of permanent jobs all over the country.[33] Similarly, with reference to transport, apartheid geographies and postapartheid underfunding of public transport combine in making traffic accidents a major cause of death (the third highest non-natural cause, at 12.5%): this meant over 14,000 traffic fatalities in 2016, the highest figure in a decade, with pedestrians nearly 40% of victims. An estimated R13 billion + productivity is lost per year in traffic jams alone.[34]

Comparative evidence suggests that MCJ's policy approach could raise standards and targets across the board—in terms of both ecology and social justice—in a way that creates incentives for both the private and public sectors to develop new technologies, infrastructure, products, services and processes, which can reduce reliance on imports and lead to economic expansion and employment gains.[35] Of course, increased regulation comes with legal and administrative costs in the short term. But when guided by public participation in its design and transparent monitoring, it promotes efficiency gains that improve productivity, and more

[31] See, e.g., AIDC (2016) and One Million Climate Jobs (n.d.).

[32] As devised by the Alternative Information and Development Centre (AIDC), part of the Climate Action Network and principal organisational advocate for the MCJ (Ashley 2018: 282).

[33] The Rhodes Biosure process, a locally developed, first-of its kind solution for treating acid mine water drainage, is arguably the most cost-effective biological treatment option developed to date for reducing sulphates in acid mine water without the external addition of chemicals (Rose 2013).

[34] BusinessTech (2017).

[35] Stefan and Paul (2008) and Epstein and Buhovac (2014).

important, increases public awareness of connections between ecological justice and social empowerment, strengthening both. Although state capacity has been retarded by billions of rands squandered on outsourced consultancy budgets, its potential is never pre-determined. Mobilizing campaigns can increase capacity to unanticipated levels; most broadly, by increasing democratic participation and decision-making, the MCJ movement would increase state legitimacy and capacity.[36]

The current challenge is one of building counter-hegemonic strength for the MCJ movement, combining a critique of the economic and ecological status quo with provisional gains wherever possible at both local and national levels. Chapter 2 detailed the ways in which the crises of "peak coal" and "peak water" coincide and reinforce each other; and Chapter 3 further discussed ways that the MEC's "peak corruption" has rendered not just the energy system, but also political legitimacy, fragile and crisis-prone. Confronting these multiply reinforcing crises with full-scale mobilization around alternatives is the best means of addressing what will hopefully be termed, retrospectively, "peak unemployment".

As Table 6.1 shows, even among the largest eight metropolitan areas, the need for immediate employment creation approaches 4 million jobs; the unemployment rate for the remaining 60% of the population living in smaller towns and rural areas is even higher. Encouragingly, however, the MCJ movement has already helped to change the terms of debate by raising the issue of unemployment to greater prominence than it has had in more than a decade, as evidenced, for example, by the convening of the 2018 Presidential Jobs Summit.[37]

To further build momentum, the goal of a "million climate jobs" could be strengthened to the level proposed by the National Development Plan 2030: one million permanent, full-time climate jobs per year over the next decade. Being class-focused without being beholden to party and union structures means it is better placed to pursue a broader vision of transformation, in terms of both national policy and energy solutions for local communities. A key strategic question is how to broaden the political coalition in favour of such a goal. While space does not permit a full and adequate discussion of this challenge, an MCJ movement could propose guaranteed minimum-wage employment to all willing to accept a climate job as a sound

[36]Carbone and Memoli (2015).

[37]Ramaphosa (2018).

Table 6.1 Unemployment per metro area, q2, 2017 (StatsSA 2017)

Metro area	Population (2017)	Unemployment rate (%)	Unemployed
Nelson Mandela Bay	1,265,000	34.5	265,650
Manguang	788,000	34.4	161,160
Ekurhuleni	3,380,000	31.2	648,960
Joburg City	4,950,000	30.1	891,000
Buffalo City	835,000	29.1	145,790
Tshwane	3,280,000	27.2	536,930
Cape Town	4,010,000	22.7	546,160
eThekweni	3,710,000	21.8	485,270
Total	22,218,000	28 (av.)	3,680,920

economic policy, no less than a means of addressing energy and climate justice goals simultaneously. Alternatively, it could focus on accelerated grid decentralization as a way of transitioning rapidly from coal. The rest of the chapter elaborates upon this dual vision by comparing full employment policy with radical grid decentralization.

JUST TRANSITIONS: COMBINING ENERGY JUSTICE WITH COMMUNITY RECONSTRUCTION

As discussed above, the MEC status quo has created simultaneous energy, fiscal, and unemployment crises. An effective opposition must highlight and draw connections among the illegitimacy of all three. Doing so requires rethinking the relationship between the national and municipal levels of government. The post-apartheid era, particularly during the Mbeki presidency, witnessed a resurgence of top-down, austerity-driven approaches to local development.[38] Indeed, it may even be an exaggeration to term this the "post-apartheid" era for local government, since

---

[38]This was first legally enshrined in the 1996 Constitution and subsequently established in the 1998 White Paper on Local Government, which introduced the notion of developmental local government, and the 2000 Local Government Municipal Systems Act, which made the activity of Integrated Development Planning compulsory for local governments. The Department of Provincial and Local Government (DPLG—renamed the Department of Cooperative Governance and Traditional Affairs in 2009) launched the Local Economic Development Fund in 1999 as part of national government's poverty alleviation strategy, leading to a proliferation of small projects, most of which collapsed once project funding dried out (Rogerson 2010: 481).

the administrative structures of municipalities were not fully integrated until 2000—massively to the benefit of the historically white (and thus wealthiest) municipalities—while the budget for local governments was slashed from R5.6 billion in 1996/7 to R1.7 billion in 1999/2000, even as local governments were burdened with a slew of new unfunded mandates, and communities with rationing via metering and user fees.[39]

These have created a chronic crisis of infrastructure and service delivery. As noted in Chapter 2, despite South Africa's middle-income status and high levels of urbanization, and despite the late-apartheid and post-apartheid governments' drive to extend grid connections to over a quarter of a million households in less than a decade, a substantial proportion of South Africans still lack access to clean, secure, and affordable energy and water. In both cases, policy makers chose to use the cumulative effects of colonial and apartheid deprivation as a yardstick for "normal" and thus "adequate" levels of consumption. Even worse, the use of metering conspired with these yardsticks to yield more regressive and burdensome results than the flat rate tariffs of the apartheid era.

An example of these dynamics can be seen in the 50 kWh monthly allocation threshold for poor households, stipulated in Section 4.1 (ii) of the Free Basic Electricity (FBE) policy of 2003. This threshold was calculated from the observation that 56% of South Africa's households at the time consumed less than this amount, on average. But since it had only been a couple of years since a bare majority of households had gained any access to electricity at all, not to mention the low levels of household wealth and income that apartheid had effected, it is not surprising that the average should have been so low. Even so, the 50 kWh monthly allocation was typically less than 10% of average household usage among township (as opposed to rural) households at the time.[40]

Similarly, infrastructure planning for the poorest households was premised on punitively low levels of consumption. In 2000, for households with an income of less than R800 per month, the Department of Provincial and Local Government (DPLG)'s Municipal Infrastructure Investment Framework supported the installation of only 5–8 Amp connections (and in newly electrified rural areas, even lower amperage of 2.5)—insufficient

[39] Gumede (2005: 78) and Hart (2012: 98–99).

[40] Fiil-Flynn and Soweto Electricity Crisis Committee (2001: 2).

power to turn on a hotplate or a single-element heater. This is less than a tenth of the standard used for most historically white areas of 60 Amps.[41]

Moreover, discriminatory and regressive tariff structures persisted in the post-apartheid era. Per-unit costs for township dwellers were typically 30% higher than for the historically white suburban areas around the time that the FBE policy was devised, and up to ten times higher than off-peak prices offered to large industrial consumers.[42] In rolling out its pre-paid meter infrastructure, the government targeted townships for their supposed "culture of non-payment" that it saw as a legacy of 1980s-era anti-apartheid boycott campaigns. In fact, large majorities of households surveyed during the Mbeki presidency had made regular contributions to their electricity bills, which they kept on file, remaining well aware of their payment situation—despite large majorities complaining that Eskom's level of service was "bad" and had deteriorated over the previous five years.[43]

Instead, non-payment was the inevitable result of the structure and implementation of the pre-paid meter policy. By setting the free allocation too low, most households' needs would inevitably exceed it. When charged at punitively high rates for excess usage, most households fell into a debt trap from which they could not escape. A 2001 survey found that 89% of Soweto households suffered electricity debt, sometimes at levels exceeding R30,000, in many cases with arrears that were more than four years old. Eskom's response was to cut off tens of thousands of households per month—a burden that a majority of the households surveyed had experienced in the preceding year—and in some cases, to permanently remove households' electricity cables as punishment for allegedly having reconnected illegally to the electricity grid.[44]

There have been over 10 million disconnections in Soweto, 60% of which were not resolved within six weeks.[45] This scale, together with the permanent removal of cables as punishment, has helped ensure that mass illegal connections would become a popular mass movement, above all in Soweto, where the Soweto Electricity Crisis Committee (SECC) and its Operation Khanyisa ("light up") emerged in 2000. The SECC demands included the

---

[41] Fiil-Flynn and Soweto Electricity Crisis Committee (2001: 5).

[42] Fiil-Flynn and Soweto Electricity Crisis Committee (2001: 1).

[43] Fiil-Flynn and Soweto Electricity Crisis Committee (2001: 1–2).

[44] Fiil-Flynn and Soweto Electricity Crisis Committee (2001: 2).

[45] Bond and Ngwane (2010).

expansion of electricity to all; free electricity; halting and reversing privatization and commercialization of electricity; and the scrapping of arrears.[46] Operation Khanyisa allowed for mass reconnections by trained informal electricians; within six months, over 3000 households had been put back on the grid. By October 2001, Eskom retreated, announcing a moratorium on cut-offs. Silumko Radebe of the Anti-Privatization Forum (APF) defended the mass illegal connections of Operation Khanyisa, arguing "the growing cost of living is now so bad that people must on a daily basis decide what they can do without. How does a person decide something like that? How do you choose between school for your kid, electricity or food in your stomach? That's a crazy choice."[47]

While these mass connections have been remarkably successful in their own right, the "moral economy" informing them points to an even more profound challenge to the political status quo. To a greater extent than most social protest movements of the post-apartheid era, such as the (tremendously successful) Treatment Action Campaign (TAC) (demanding universal provision of anti-retroviral drugs for all HIV-positive South Africans), or the Fees Must Fall students' movement (demanding tuition-free access to university), as well as protests against road tolls, Operation Khanyisa succeeded in making lateral, solidaristic linkages among community groups to challenge the underlying logic of austerity, and not just its particular reapportionment. As an ANC official grudgingly observed in 2003, community mobilizations against electricity cuts have expanded their scope of demands to include

> a demand for houses, a stop to eviction/relocation, and access to free basic water among other issues. This is essentially a call to develop a broad united front that goes beyond [electricity]. ... [The APF] also creates the imperative link between the shop floor struggles against right-sizing (retrenchments), casualization of labour, and the struggles waged against water and electricity cuts in the townships. ... It is this ability to link these cuts of services and electricity to privatization that creates a strong and broader appeal.[48]

The local mobilizations and issue linkages also helped Sowetans avoid the fate of their poor Cape Town counterparts and reject pre-paid water meters

---

[46] Bond and Ngwane (2010: 198).

[47] Styan (2015: 97).

[48] Quoted in Bond and Ngwane (2010: 203).

and rationing. But these gains have yet to become universalized as formal changes of policy.

Ironically, Eskom's corruption, on top of practical difficulties, has prevented municipalities from using cutoffs as punishment as they are permitted to do under the 2000 Promotion of Administrative Justice Act.[49] Although municipalities charge an extra percentage point above the Eskom tariff rate—with the ostensible purpose of paying for grid maintenance, but increasingly, to balance their budgets—they are increasingly in debt to Eskom, in aggregate, for more than R15 billion (with Soweto, for roughly R10 billion), not withstanding generally dubious accounting. The Treasury, meanwhile, has refused Eskom's request to assume R100 billion of its debt.

Yet even while anti-austerity movements partially succeeded in thwarting the government's drive to privatize basic services, the imposition of austerity budgets upon municipalities remains. Successfully challenging austerity entails challenging the legal framework and underlying assumptions that inform it. The 2000 Act, for example, is consonant with the DPLG's National Framework for Local Economic Development, a set of guidelines for local authorities to pursue local economic development policy and practice, by using "local competitive advantage" with the aim of creating "robust and inclusive local economies, exploiting local opportunities, real potential and competitive advantages, addressing local needs and contributing to national development objectives."[50] Chronic underfunding further exacerbates previous problems, such as duplication between the activities and responsibilities of the two major central government line ministries (the DPLG and the Department of Trade and Industry [DTI]).

In other words, the climate jobs movement can argue forcefully that just as coal-powered electricity is ecologically unsustainable, and "tenderpreneurship" is politically unsustainable, so austerity budgeting and basic water and electricity metering is both economically unsustainable and morally unjustifiable. A continuation of the REIPPPP addresses the ecological critique imperfectly at best, but leaves the other two unaddressed. Below, among many possibilities, two alternative approaches to combining all three critiques are discussed. The first, dependent on a sufficient degree of centralised government coordination with provincial and municipal counterparts, proposes guaranteed climate jobs employment for all

---

[49] Styan (2015: 104).

[50] DPLG (2006: 17).

willing to take these jobs. While addressing decarbonisation and promoting community-led green development, this policy prioritises the rapid reduction and eventual elimination of structural unemployment. The second approach more sceptically assumes less capacity for central-local government coordination and prioritises rapid energy transition and decarbonisation. It too would entail substantial unemployment reduction—generating hundreds of thousands if not millions of jobs per year—but would leave the impetus for job creation primarily at the provincial level, while replacing Eskom's centralised generation and transmission of electricity with a much more decentralised alternative.

Guaranteed Full Employment: Achieving Economic Recovery with Climate Jobs

In his first State of the Nation address in 2019, President Ramaphosa listed the first task of government as being to "accelerate inclusive economic growth and create jobs." He acknowledged that the "levels of growth that we need to make significant gains in job creation will not be possible without massive new investment." He cited the previous year's Presidential Jobs Summit whose proposed measures would add 275,000 jobs and thereby nearly double the current number of jobs created annually (about 300,000).[51] At under 600,000, this is 25% fewer than the number of matriculating students per year. Clearly, this rate is far too slow: even previous periods of higher growth never created jobs at the rate needed—much less for the populations in greatest need. The Job Summit target is less ambitious than the country's 2010 National Development Plan (NDP) goal of 11 million jobs cumulatively by 2030, to reduce unemployment to 6%—and structural unemployment has steadily worsened in the decade since the NDP was drafted. In the unlikely scenario that fully half of all the proposed news jobs went to the NEET (not in employment, education or training) youth alone, it would take more than fifteen years to seriously tackle their unemployment—and much longer to bring total unemployment below 6%. Moreover, South Africa presently has no prospect of sustained levels of GDP growth: annual real GDP per capita growth has remained below 5% for over a decade. Excessive trade exposure and currency liberalisation has severely damaged the country's

---

[51] Tshwane (2018).

manufacturing base, making it unlikely to compete globally in any labour-intensive sector. As President Ramaphosa admitted in his address, the country has recently slipped into recession.

How then can the rate of employment creation be accelerated?[52] As noted above, most standard employment creation policies face lag times and need to be tailored to specific circumstances,[53] particularly if disadvantaging poorer and more marginalized workers is to be avoided.[54] The urgency of providing work for millions of the unemployed, however, is more immediate than the time it takes for most of these measures to take full effect.

A policy that can take immediate effect is directly subsidized employment. Building on the pioneering work of economist Hyman Minsky, this is a policy that has already been shown to work over the short-term in the Global South. In Argentina, the *Jefes y Jefas de Hogar Desocupados* (Program for Unemployed Male and Female Heads of Households) was introduced after that country's financial crisis in 2001. Federally funded and locally administered, the program offered guaranteed employment of at least four hours a day in community-created jobs to the unemployed heads of households. It showed success in reducing poverty as well as feelings of powerlessness and social marginalization, especially among its (majority) female participants, but was phased out after a few years and replaced with more traditional social spending efforts.[55] More recently, in India, the 2005 National Rural Employment Guarantee gives up to 100 days of guaranteed paid employment per year to workers from rural households. Although implementation varies regionally, reported results include 13% increased earnings in low-income households and a reduced gender pay gap. Direct benefits, including increases in wage income, exceeded program-related transfers, especially for some of the most socially marginalised groups: government-targeted castes and tribes and households supplying casual labour. Women

---

[52] This section draws on Lawrence (2019).

[53] Braunstein and Heintz (2008).

[54] The experience of India for example shows that deflationary policies disfavour both women and the rural poor, and by extension, agricultural productivity. When the government increased expenditures on rural development and employment generation, women's income and agricultural productivity improved; when these budgets were cut, women's income and agricultural productivity declined and infant mortality sharply increased (Patnaik 2003).

[55] Tcherneva and Wray (2005) and Garzón de la Roza (2006).

in particular also reported benefitting through increased food security and being more able to avoid more hazardous work.[56]

What would happen if the state were to guarantee full-time jobs to all South African workers willing to take them? In this scenario, the state guarantees employment at the national minimum wage, paying R3500 per month, and works with provinces and municipalities to allocate openings in the areas of mutually agreed-upon highest priority. Because this wage is itself higher than the wages of 47% of working-age South Africans, such a guarantee could help to reduce income inequality.[57] Guaranteeing full-time work for the 6 million officially unemployed would cost R252 billion; for the 9 million workers including those currently discouraged from seeking work or in unstable or part-time work, it would cost nearly R380 billion.

Fiscal conservatives would argue that South Africa cannot afford an additional welfare provision at this cost; R380 billion is more than twice the amount spent on the country's most important welfare provision, its Social Grants programme, which benefit one third of the population—more than 17 million people, including 11 million children—and cost around R150 billion a year. In his inaugural address, President Ramaphosa similarly implied that "spending our way out of our economic troubles" would be somehow at odds with "setting the economy on a path of recovery"—a veiled reference, perhaps, to the hundreds of billions of rands of state-capture largesse. Full employment advocates could also point to the hundreds of billions per year that fossil fuel subsidies and load-shedding have cost the country over the past decade (as detailed above) and argue that climate jobs could contribute to an accelerated energy transition. Measures targeting capital flight could also fund the program's inauguration.

There are, moreover, strong reasons to think that to "spend our way out of our economic troubles" by focusing on labour-intensive employment creation may in fact be among the fastest and most durable ways of "setting the economy on a path of recovery." The reason (as noted in Chapter 2) is that reducing unemployment has a positive, direct, short-term effect on growth: over one percentage per year for every 5% unemployment reduction, by some estimates.[58] Guaranteeing full-time minimum-wage employment for five percentage points of the unemployed—around

---

[56]Deininger and Liu (2013) and Khera and Nayak (2009).

[57]Valodia and Francis (2016).

[58]See SAMI (2018).

465,000 workers—would cost R19.53 billion. Yet this amount is little more than one third of 1% of GDP (about R55 billion). And with each reduction of five percentage points of unemployment in this model, GDP increases by about R550 million.

There are several reasons why labour-intensive policies leading to immediate decreases in unemployment may have a significant and positive knock-on effect on growth. A widely accepted Keynesian perspective on labour markets understands that labour markets do not clear in the way that neo-classical models posit and hence recognises involuntary unemployment. But post-Keynesian perspectives relevant to South Africa's massive levels of structural unemployment add an additional insight.[59] Under currently depressed conditions, South African firms have unused capacity (about one fifth).[60] Because workers have a higher propensity to spend their income than do capitalists, an increase in labour's share of income boosts aggregate demand—to which firms would immediately respond by increasing output rather than prices. Because low-wage workers spend a far higher proportion of their income on locally produced goods than their higher wage counterparts, this money would circulate further in the national economy than is the case with the type of large capital expenditures that President Ramaphosa may be referring to—and which South Africa has an unfortunately extensive track record in making.[61] Once incentivized to invest, South African firms bring capacity utilization back towards its normal level. In other words, there is a positive relationship between the wage share and output, with increasing wage share leading not only to higher consumer demand, but also to higher investment and capital stock expansion.[62] Reducing unemployment to 2% could therefore increase GDP by five percentage points—to a level higher than any in the past decade littered with megaprojects costing hundreds of billions of rands, and exceeded in only four years since 1994.

---

[59] Blecker (2002).

[60] Laing (2018).

[61] In addition to the $53-billion rail line built to export billion tons of coal on South China Rail locomotives, further examples include the Medupi and Kusile coal-fired power stations costing $15 billion each; and the $17-billion Durban port-petrochemical expansion; the Coega complex (R20 billion), the World Cup stadiums (R27 billion), and the 1999 arms deal (R43 billion)—together costing R1.5 trillion (Bond 2019).

[62] Stockhammer (2015).

There are at least three additional arguments in favour of a full-employment policy. The first, as observed earlier in the chapter, is that not only poverty, but also violent crime, is directly correlated with unemployment. When unemployment goes down, so does the cost of crime to society as a whole. The second reason is that a full-employment policy would give a powerful impetus to reducing inequality. This too would have a direct effect on reducing crime's R1 trillion tax on the country. The third, and strongest, reason for full employment is economic justice. Since climate jobs would be strong contenders for the type of employment generated, economic, energy and climate justice could all be realised simultaneously.

Given the fluidity of the current state of South African politics, there is a chance that the current government formed by the ANC will be more amenable to adopting bolder policy measures, such as guaranteed full employment, than its predecessors. As noted, however, this policy depends on the central government effectively coordinating with provinces and municipalities to agree upon, and implement, areas of employment priority. The capacity of local government to conduct effective audits and water monitoring appears to correlate substantially with their average income. Although only policy experimentation can make an accurate assessment in this regard, it may be the case that, initially at least, the local governments in greatest need of employment creation are those least able to manage its implementation.

Under a more pessimistic scenario in which either national political forces advocating full employment are insufficiently strong or local capacity to implement full employment is inadequate (or both), advocates for just transitions and climate jobs may need to adopt a more ad hoc approach to advocacy and employment creation until the national political context becomes more favourable. Some climate jobs sectors—such as mass public transport, including heavy and light rail expansion—enjoy sufficient national oversight that local capacity is less of an issue. As noted, public transport could substantially reduce road and pollution-related deaths as well as emissions and introduce infrastructure to underserved and economically depressed areas, such as between East London and Durban or Tshwane to Musina.

Given the protracted crisis of service delivery at Eskom, however, accelerated decentralization of generation and replacement of coal with alternatives has become a matter of urgency for national energy security. Relying on the REIPPPP to achieve this risks increased indebtedness and financial exposure, adding capacity too slowly, ignoring labour movement and civil society demands for energy decommodification while having a grossly inadequate effect on employment creation. Therefore, a different mechanism for scaling up RE generation while increasing employment-led and demand-led growth may be necessary.

A People's REFIT: A Decentralised Pathway to Energy Justice?

The political weakness and dependence of municipalities upon the central state require both democratisation and a degree of ownership and control of sources of income. Unbundling Eskom, as proposed by the 2011 Independent Systems and Market Operator Bill, seems increasingly likely to be adopted, and could lead to energy privatization; but it could also serve progressive interests at the local level as well. Climate jobs advocates could demand that the grid become owned and controlled at the provincial or local levels, with the power to set levies for its upgrading and maintenance. In exchange for municipal debt-forgiveness, municipalities in turn would adopt a people's REFIT with a progressive tariff structure, as has been adopted in countries ranging from Bulgaria to Vietnam. As discussed in Chapter 3, several factors—lack of ministerial coordination, mega-coal fired generator lock-in, state capture, and a lack of transparency and accountability at Eskom—combined to jettison the 2009 REFIT policy before it had been adequately considered, let alone implemented. This episode (and the MCJ campaign launch) coincided with the 2011 United Nations Climate Change Conference (COP17) in Durban. Although South Africa's green and trade union civil society groups did not strongly embrace the REFIT policy at the time, given the increasing social, political and economic costs of coal and decreasing costs of RE sources since then, the case for doing so has arguably grown even stronger. As discussed in the previous chapter, their overriding concerns are that the energy infrastructure be nationally coordinated, with decentralised social ownership improving on current levels of service provision and satisfaction of universal needs, while providing communities and workers with an equitable dividend in energy's production and consumption.

As the experience of dozens of states and jurisdictions around the world has shown, feed in tariffs (FITs) are among the best policy instruments for attaining these several goals simultaneously. By requiring Eskom to purchase all surplus-generated electricity from all eligible producers over an extended period of time (typically, 15–20 years), as well as to connect these producers to the central grid and cover grid expansion or interconnection costs, FITs send a strong signal to potential and actual producers that their returns on investment are guaranteed for this timeframe. With financing assistance from an enabling state, communities can then organize producer or worker-owned cooperatives and/or impose shared ownership and control requirements to utility-scale providers who wish to make use of their sites (rooftops, farmland, etc.), benefiting directly from income generation. A progressive tariff structure, targeted according to neighbourhoods' or households' average income, generation level, region, and/or to sector (commercial, industrial, residential, agricultural, or extractive—thus also potentially directed at the main coal mining regions), would reverse the current legacy of regressive rates. It could additionally privilege storage infrastructure, such as an expansion of the already-existing pumped storage system. Together with grassroots participation in its implementation, progressive, pro-poor tariffs should be a central element of any FIT; they could serve as arguably the first truly broad Broad-Based Black Economic Empowerment policy, and without the BEE-levels of red tape.

The beneficiaries could include not only the 85% of South Africans who currently enjoy at least nominal grid access: a progressive FIT would also serve as arguably the best and fastest incentive for many of the remaining 15% to lobby for and organize access, or subsidise the creation of mini-grids in remote rural areas.[63] In addition, it would benefit all consumers with a vastly more secure energy supply (with hundreds of thousands, if not millions, of independent producers rather than a few power plants). This would both sharply reduce the risk of blackouts and the need for scheduled load-shedding, and thereby result in overall electricity cost reductions. As underscored previously, an assessment of the cost of an FIT along these lines should include the many hidden tax, load-shedding, state-capture, and health and environmental opportunity costs of the status quo. It should also recognise that as a progressive tax, such an FIT would help to redress

---

[63] By comparison, in Germany over the past decade, between 40–50% of all citizens (almost 50% in 2012) owned the country's renewable energy capacity through energy cooperatives and private initiatives, including nearly half of all installed biogas and solar capacity and half of the installed onshore wind capacity (Amelang 2016; Borchert 2015; Nestle 2014). With higher levels of poverty and unemployment, the incentives for a larger proportion of South Africans to participate in similar cooperative arrangements would be greater.

the regressive structure of electricity tariffs imposed upon the poor and working class majority over the past two decades.

There are broader fiscal dimensions to consider as well. Eskom is currently massively indebted to the tune of more than R350 billion; given debt repayments that will soon exceed R50 billion per year, the total amount is likely to exceed half a trillion rands. The question of who should pay for this debt in principle is little different from that of who should pay for the debt incurred under apartheid; both merit a full audit. In the narrowest sense, the democratic government inherited from the apartheid regime a banking and private sector debt of $10 billion, and a public sector debt of $15 billion. The 1998 Apartheid Debt and Reparations Campaign has called for cancellation of this debt, filing legal suits against the major transnational corporate and financial interests that benefitted from apartheid on behalf of victims throughout southern Africa.[64] At the time, *The Economist* argued that "it is not obvious how such a plan for reparations might be implemented fairly. Were blacks harmed, for instance, by foreigners who invested in South Africa's public electricity firm?"[65] But by now the answer to this rhetorical question should be obvious. To these odious debts should now be added those associated with Medupe and Kusile, whose completion is predictably massively over-schedule, technically unlikely, unnecessary, and fiscally, ecologically and socially detrimental.

Because the FITs would be universally applied, they would impose a level of generation and transmission transparency upon Eskom (or its successor entities) greater than it has ever shown. Since its net cost effect could be progressive, with properly structured incentives for both households and municipalities, the largest electricity consumers would have the strongest incentives to increase their energy efficiency levels. Eskom would benefit from displaced fuel costs, and vehicle fleet electrification would benefit small businesses and consumers as well: at present, petroleum (both crude and refined) represents the largest category of imports, costing over $9 billion in 2017.[66]

[64] IADRC (2003).

[65] Economist (1999).

[66] Two of South Africa's top imports are Crude Petroleum ($6.54 billion) and Refined Petroleum ($2.55 billion) (OEC 2018).

Eskom and consumers would also benefit from decreased volatility of fuel and electricity costs.[67] An additional, particular windfall to Eskom is that the current costs associated with degradation of infrastructure (e.g. copper cable theft, in 2018 estimated to cost R5–7 billion a year)[68] and the understandable refusal of poor consumers to pay for expensive, unfairly assessed and unreliable power could also be vastly reduced, if not dispensed with entirely.[69] As net FIT beneficiaries, township residents and the poor would come to see the power infrastructure as their own, and a major source of household revenue, rather than view it as the hostile and alien presence it has been up to now. South Africa's inherited geography of (neo)apartheid could be transformed. Income could be generated far more locally than at present, and more small businesses could proliferate. Traffic congestion could decrease and neighbourhood cohesion increase.

Rooftop PV conversion alone could have a far greater employment effect than the entire REIPPPP thus far. According to IEP figures, onshore wind yields 6401 job years/GW, and solar PV, 8724/GW, or 15,125 for capacity expenditure plus 11,350 for operational expenditure, yielding 26,475 jobs per 100 TWh/yr—30% more than new coal yields to produce the same amount of energy. An accelerated decommissioning of coal would therefore *increase* total employment in current coal mining regions if these were prioritized.[70] These outcomes could be achieved if a progressive FIT were implemented together with universal net metering (NM). NM enables grid-connected renewable systems to be credited for the electricity that they provide to the grid. It can, and often has, complemented FIT goals. Unlike double-metering, which often entails dual pricing that discriminates again RE sources and always includes exorbitant transaction costs (verifying generation, separate payment systems, etc.), net metering simply pays the going rate for surplus generation. It also entails an intrinsic incentive to save as well as produce energy when prices are higher and consume when they are lower. A given household or community cooperative would become not merely consumers, but "prosumers" (producer/consumers) of electricity—and since the majority currently consume comparatively little, their net earning potential is greatest.

[67] Mendonça et al. (2009: xxi).

[68] Theron (2018).

[69] Omarjee (2018).

[70] DoE (2016). IEP Annexure B: Macroeconomic Assumptions.

As an intrinsic source of competition to a utility provider such as Eskom, NM is often resisted or opposed by utilities. Indeed, Eskom currently discourages small scale RE customers from producing more energy during a year than they consume. Among other downsides, this stricture exacerbates apartheid geographies and income disparities, since it places the lowest cap on the poorest households rather than use NM as a means of income generation and redistribution. Moreover, comparative studies show that the most successful NM policies are those that impose no limits on maximum system capacity, or restrictions on RE sources, require all utilities to participate and include all types of customers (households, small and large businesses, etc.), without additional charges, application fees or tariffs.[71]

To be sure, universal net metering is the indispensable precondition for a "people's FIT" with a progressive, rather than regressive tariff structure, to work. As a technical barrier, it is trivially easy to implement. Indeed, in key respects, Eskom's current policy of tolerating but not coordinating a haphazard number of NM households represents the worst of both worlds. As Tobias Bischof-Niemz (until recently head of the CSIR Energy Centre) has argued, if one million higher-income households were each to install 6 kWp of PV, together with another 20,000 commercial properties installing 200 kWp each, this would represent a total of 10 GWp of installed PV capacity "under the radar" of Eskom's capacity to monitor and coordinate. This would impose serious safety concerns for grid operators and line workers confronting potentially large numbers of technically non-compliant PV installations; moreover, without a commonly agreed tariff policy, municipalities (currently already massively in the red) risk future bankruptcy. For this reason, South African Local Government Association (SALGA) is developing a net metering proposal providing structure to municipalities' existing embedded PV customers, pending the final drafting and approval of a wiring code for embedded generators (South African National Standards, SANS 10142-3) by the South African Bureau of Standards (SABS) that would instruct electricians about safe procedures for installing embedded generators.[72] Several municipalities have advanced plans to implement, or have already begun implementing, net metering schemes, including a majority in the Western Cape.[73]

---

[71] Mendonça et al. (2009: 164–166).

[72] SALGA (n.d.).

[73] Western Cape Province (2019) and Bischof-Niemz (2015: 18).

CSIR has proposed a master plan combining NM and FIT policies for the whole country—the NETFIT—via a "Central Power Purchasing Agency (CPPA)". This would provide a FIT for all surplus generated energy, while compensating municipalities and Eskom Distribution for reduced sales resulting from households' self-generated energy consumption. With a total annual funding requirement of R530 million for every 500 MWp[74] of embedded PV built under the scheme, the net cost is projected to be a mere R290 million per year for every 500 MWp of PV capacity it installs, which translates into an increase of the average tariff of ± 0.15 R-ct/kWh (an already low amount that would decrease over time virtually to zero as more and more capacity is added).[75]

Again, a sufficiently progressive FIT could ensure that the poorest two-thirds of households would be net beneficiaries and thus unaffected by tariff increases—and indeed, given the extensive list of fraud, waste and abuse detailed earlier in the chapter, addressing these could render tariff increases unnecessary. Additionally, since (as noted above) cable theft and non-payment cost over R20 billion per year, and net metering provides a clear incentive to end these practices, they alone could pay for more than 30,000 MWp (or 30 GWp; at 3 h per day, 90 GWh/day, 32,850 GWh/year, or 3.75 GW)—more than the country's entire PV capacity in 2016, and about twice the capacity of the latest REIPPPP bid round. By comparison, industry association SAPVIA has advocated 1.5 GW of PV per year for the next five years.[76]

Even though there is currently a surplus of generating capacity, there are strong arguments for massively expanding RE generating capacity in the immediate term. First, the reliability of the grid would be strengthened, and risks of further blackouts thereby reduced. Indeed, a sufficient degree of surplus RE capacity would render the IRP, which has proven prone to gaming and political manipulation, largely obsolete, since small-scale generation could be added onto or taken off the grid within a matter of days or weeks; the main policy decision would entail adjusting tariff rates. Second, decommissioning of coal-fired plants could be accelerated, and thus air quality improved more quickly; and third, there remains vast potential to further

---

[74] Abbreviation for megawatt peak, a measuring unit for the maximum output of an intermittent RE source (e.g. a pv plant). One MWp = 1000 kilowatt peak (kWp). 500 MWp would generate about 1–2 GWh per day in SA.

[75] Bischof-Niemz (2015: 27).

[76] Breytenbach (2017).

electrify South Africa's industrial infrastructure, most obviously, its vehicle fleet. Although vehicle electrification is still in its infancy in South Africa, with declining vehicle costs (from combis and minibuses to automobiles to scooters) and even more swiftly declining electricity costs compared to petrol—and in the wake of state-owned Chinese automaker BAIC Group's recently announced plans to produce electric vehicles in South Africa—it may be only a matter of a decade or two before they become the norm. Even at current prices, EV charging costs are lower than petrol per 100 km travelled.[77]

The main inhibition is currently the absence of a national charging infrastructure. Encouraging local or municipal ownership of infrastructure should be straightforward. On the equivalent area of fifteen to thirty residential roofs, 950 sq m of rooftop photovoltaic (PV) panels can generate up to 1.9 GWh per year (roughly equivalent to one 1 MW-capacity onshore wind turbine).[78] This amount of electricity is enough to power these thirty average-sized houses plus one five-port electric vehicle (EV) charging station.[79] The construction and maintenance of one five-port station could generate at least fifteen jobs. With about 5000 petrol stations in South Africa currently, this could scale up to over 75,000 jobs a year—about as many as are currently employed in coal mining. Each station would also reduce up to 700 metric tons of emissions per year through avoided coal generation, not including 250 tons of avoided petrol emissions; together, saving thousands of lives and millions of rands in health costs by securing cleaner air.

Without a progressive FIT complement, NM arguably does not adequately incentivize localized RE deployment, nor does its pricing mechanism fully reflect the (health, employment, etc.) benefits of RE sources over alternatives such as coal. The NETFIT proposal includes the crucial additional provisions of rendering investments secure by guaranteeing export (i.e. surplus-generated) tariffs over the lifetime of the RE asset, without

[77] In 2017, a 50-litre tank with 93 unleaded petrol inland (including tariff costs, discounts, et al.) cost R654 (Wheels24 2017). Assuming 16 km a litre yields a range of 800 km per fill up. The average EV requires 20 kWh to fully recharge; at the current rate of 120 cents per kWh, it would cost R24 to fully recharge ("fill up") the car. A 2017 Ford Focus Electric has a range of 185 km. Thus, 4.3 recharges = 1 full petrol tank, but cost only R103; thus petrol costs more than six times the cost of electric recharges per km.

[78] Assuming 1 sq m of PV generating 2 MWh per year, times 950 = 1.9 GW.

[79] Energati (2016).

imposing fixed charges per household; fostering the rise of a micro-utility small businesses or energy cooperatives; creating a nationally centralized off-take of generated surplus; and providing funding sources external to municipality budgets, rather than dependent on them.[80] Conversely, without a participatory governance component, NM is unlikely to be able to fully mobilize communities to use RE as a means to socio-economic empowerment. Both are needed.

Conclusions

This chapter has contrasted three contingent scenarios: the business-as-usual (BAU) scenario that continues the REIPPPP, subordinate to the interests of the transnational mining and coal sector; a guaranteed employment policy as a means of scaling up climate jobs; and the NETFIT scenario that rapidly scales up small-scale RE onto the decentralized grid, via net metering and a progressive FIT. Although by no means the only options available for energy reform, it is worth elaborating upon these scenarios in order to explicitly distinguish between what is more and what is less certain; what is more and less amenable to change over the short- to medium-term; and between what is normatively desirable, and what entails unavoidable trade-offs or uncertain costs.

The BAU scenario faces the certainty of a protracted electricity generation and load-shedding crisis at Eskom in the immediate term, and growing externalities (in terms of stagnant employment growth, growing health costs, growing water shortages and water and air pollution, high, chronic levels of structural unemployment and inequality, and ensuing from these, extremely high costs of crime and violence prevention) with each additional year. It also faces the certainty of fiscal and macroeconomic imbalances resulting from excessive dependence on the mining sector, as well as of the eventual decline of coal as both a viable export commodity and domestic energy source. South Africa's dependency on coal is drawn out for as long as possible to the bitter, polluted and even more structurally unemployed end, while leaving the country more bereft of a Plan B and with each passing year, ever more vulnerable to increasingly extreme cycles of ever-shorter mineral commodity booms followed by ever-longer busts, ever-shorter (but

---

[80] Bischof-Niemz (2015: 27).

increasingly devastating) floods and ever-longer droughts, and thus growing macroeconomic and human security fragility.

What is less certain is precisely how government and the organized interests of mining and unions will respond to the urgently necessary decision to replace coal. Although coal sector job losses are inevitable, the question of *if and how* they are planned for (or not) makes a huge difference to workers' lives. NUM and other key unions in coal mining need to have an active, official role in both avoiding involuntary unemployment during the transition (through early retirement, workweek reductions, etc.) and in tailoring new RE-related work programs locally.

It is also less clear how much more mileage the REIPPPP has under its current modus operandi. On the one hand, it would need to be replicated at a multiple of several times the current generating capacity in order to make substantial inroads against coal; on the other, the provisions for community employment and empowerment seem increasingly burdensome to bidders, as well as ineffective to inadequate by civil society groups and communities. Most important, its financing model seems politically and fiscally unsustainable. The first three bidding rounds may seem in hindsight like low-hanging fruit. Different approaches to combining the goals of cleaner, more secure energy provision with localized development are needed. Advocates of a focus on RE generation to the subordination of job creation and community empowerment and engagement need to recognise that this too comes with significant political costs.

Regarding unavoidable trade-offs and uncertain costs, advocates of a continued reliance on coal and mining more generally cannot ignore the trade-off choice of paying for social, economic, and environmental externalities proactively in order to reduce larger costs later on. There is the additional unavoidable trade-off between paying for universal net-metering up front, or suffering the consequences in terms of increased grid repair hazards and risk of municipal bankruptcy that would accompany the status quo of haphazard net-metering. The economic costs, as we have seen, of the current regressive electricity tariff structure and punitive service cut-offs that penalize low-income households include over R20 billion a year due to (unevenly accessed) electricity commoning and cable theft. A less certain, but potentially immense cost, is that of mistrust in, and cynicism about, representative government in general and its parastatals in particular.

Regarding what is more and what is less certain, and more and less amenable to change over the short- to medium-term with the guaranteed employment proposal, the crisis of unemployment as already noted has attained greater political traction than arguably ever before. Unions and community organisations can only embrace the basic thrust of this proposal. The key is to convince policymakers and business interests that it is in their enlightened self-interest, as it could attain higher levels of GDP growth and lower levels of violence and crime. The capacity of municipalities to absorb job seekers at sufficient scale is also untried and thus unknown; but early success in more readily monitored sectors could increase capacity in this regard.

Regarding what is more and what is less certain or amenable to change with the NETFIT proposal, it is certain that its adoption would meet with stiff resistance from Eskom and from the coal industry interests. It is also certain, for this reason, that a broad coalition of latent beneficiaries—among consumers, civil society, municipalities, labour and the private sector—would therefore need to lobby strenuously to get it accepted. It is likely that opponents would be less opposed to a universal rollout of net metering (which would address the grid safety concerns mentioned above) than to the progressive FIT component. It may be that these two stages would need to occur sequentially, even though their immediate adoption would accelerate a RE transition. What is less certain is what its full potential might be regarding reducing inequality and promoting decentralized employment. This depends on the extent to which and speed with which communities that are not grid connected can become connected, which in turn probably depends on how strong the prior lobbying campaign can be, and on the level at which the FIT is set.

From the perspective of the MCJ campaign, three types of event may prove essential for its success. The first is the unceasing stream of failures—revelations of corruption, breakdowns, load-shedding, and persistent lack of adequate electricity access—of the status quo. Each occasion becomes an opportunity to provide persistent reminders that a better alternative is currently available. The second is the increasingly extreme recurrence of droughts, floods, and gales sweeping the land. Not only are they symptoms of what is wrong with the status quo, but they are occasions for the immediate provision of alternative infrastructures, jobs, and policies. The third is the actual achievement of these alternatives. Nothing succeeds like success, and a growing consciousness of alternatives is already evident.

References

Alda, E., & Cuesta, J. (2011). A comprehensive estimation of costs of crime in South Africa and its implications for effective policy making. *Journal of International Development, 23*(7), 926–935.

AIDC. (2016). *One million climate jobs: Moving South Africa forward on a low-carbon, wage-led, and sustainable path.* Accessible at: http://aidc.org.za/download/climate-change/OMCJ-booklet-AIDC-electronic-version.pdf.

Alfreds, D. (2018, April 13). Rural North West school gets green electricity. *News24.* Accessible at: https://www.news24.com/Green/News/rural-north-west-school-gets-green-electricity-20180413.

Amelang, S. (2016, June 29). Germany's energy transition revamp stirs controversy over speed, participation. *Clean Energy Wire.* Accessible at: https://www.cleanenergywire.org/dossiers/reform-renewable-energy-act.

Ashley, B. (2018). Climate jobs at two minutes to midnight. In V. Satgar (Ed.), *The climate crisis: South African and global democratic eco-socialist alternatives* (pp. 272–292). Johannesburg: Wits University Press.

BBC. (2018a, August 14). South Africa to eradicate pit latrine toilets in schools. *BBC News.* Accessible at: https://www.bbc.co.uk/news/world-africa-45183593.

BBC. (2018b, July 6). South African boy of three drowns in Limpopo toilet. *BBC News.* Accessible at: https://www.bbc.co.uk/news/world-africa-44736772.

BBC. (2018c, September 18). South Africa crime: Can the country be compared to a 'war zone'? *BBC News.* Accessible at: https://www.bbc.co.uk/news/world-africa-45547975.

Bischof-Niemz, T. (2015, June). *How to stimulate the South African rooftop PV market without putting municipalities' financial stability at risk: A Net Feed-in Tariff proposal.* CSIR/02400/Renewable Energy/IR/2015/0002/A. Accessible at: http://www.wind-works.org/cms/fileadmin/user_upload/Files/Reports/South_Africa_NETFIT_concept_-_CSIR_-_8Jun2015.pdf.

Blecker, R. (2002). Kaleckian macro models for open economies. In J. Deprez & J. Harvey (Eds.), *Foundations of international economics: Post-Keynesian perspectives* (pp. 126–160). London: Routledge.

Bond, P., & Ngwane, T. (2010). Community resistance to energy privatisation in South Africa. In K. Abramsky (Ed.), *Sparking a worldwide energy revolution: Social struggles in the transition to a post-petrol world* (Chapter 15, pp. 197–207). Oakland, CA: AK Press.

Borchert, L. (2015, March 10). Germany between citizens' energy and Nimbyism. *Clean Energy Wire.* Accessible at: https://www.cleanenergywire.org/dossiers/peoples-energiewende.

Braunstein, E., & Heintz, J. (2008). Gender bias and central bank policy: Employment and inflation reduction. *International Review of Applied Economics, 22*(2), 173–186.

Breytenbach, M. (2017, December 8). SAPVIA launches 'bold' 5-point plan for solar industrialization. *Creamer Media's Engineering News*. Accessible at: http://www.engineeringnews.co.za/article/sapvia-launches-bold-5-point-plan-for-solar-industrialisation-2017-12-08.

Budlender, D., Casale, D., & Valodia, I. (2010). Gender equality and taxation in South Africa. In C. Grown & I. Valodia (Eds.), *Taxation and gender equity: A comparative analysis of direct and indirect taxes in developing and developed countries* (pp. 206–232). New York: Routledge.

Burton, J., Lott, T., & Rennkamp, B. (2018). Sustaining carbon lock-in: Fossil fuel subsidies in South Africa. In J. Skovgaard & H. van Asselt (Eds.), *The politics of fossil fuel subsidies and their reform* (pp. 229–245). New York: Cambridge University Press.

BusinessTech. (2016, June 10). How much money violent crime costs South Africa every year. *BusinessTech*. Accessible at: https://businesstech.co.za/news/general/126263/how-much-money-violent-crime-costs-south-africa-every-year/.

BusinessTech. (2017, June 9). South Africa's shocking road death numbers at highest level in 10 years. *BusinessTech*. Accessible at: https://businesstech.co.za/news/motoring/178275/south-africas-shocking-road-death-numbers-at-highest-level-in-10-years/.

Carbone, G., & Memoli, V. (2015). Does democratization foster state consolidation? Democratic rule, political order, and administrative capacity. *Governance, 28*(1), 5–24.

Deininger, K., & Liu, Y. (2013). *Welfare and poverty impacts of India's national rural employment guarantee scheme: Evidence from Andhra Pradesh*. Washington, DC: The World Bank.

Department of Energy (DoE). (2016). *Integrated energy plan annexure B: Macroeconomic assumptions*. Accessible at: http://www.energy.gov.za/files/IEP/2016/IEP-AnnexureB-Macroeconomic-Assumptions.pdf.

DPLG (Department of Provincial and Local Government). (2006). *Stimulating and Developing Sustainable Local Economies: National Framework for Local Economic Development (LED) in South Africa*. Pretoria: DPLG.

Economist. (1999, April 22). South Africa's debt: Unforgivable. *Economist*. Accessible at: https://www.economist.com/finance-and-economics/1999/04/22/unforgivable.

ENCA. (2015, June 5). How does biogas technology work? *ENCA*. Accessible at: https://www.enca.com/south-africa/gallery-how-does-biogas-technology-work.

Energati. (2016, January 19). South Africa's solar market needs innovative financing. *Energati*. Accessible at: http://www.engerati.com/article/south-africa%E2%80%99s-solar-market-needs-innovative-financing.

Epstein, M. J., & Buhovac, A. R. (2014). *Making sustainability work: Best practices in managing and measuring corporate social, environmental, and economic impacts*. San Francisco: Berrett-Koehler Publishers.

Fajnzylber, D., Lederman, D., & Loayza, N. (2002, April). Inequality and violent crime. *Journal of Law and Economics, 45*, 1–40.

Fiil-Flynn, M., & Soweto Electricity Crisis Committee. (2001). *Electricity crisis in Soweto* (Municipal Services Project Occasional Papers No. 4), 1.

Foster, N., Vassall, A., Cleary, S., Cunnama, L., Churchyard, G., & Sinanovic, E. (2015). The economic burden of TB diagnosis and treatment in South Africa. *Social Science & Medicine, 130*, 42–50.

Freeman, R. B. (1992). Labor market institutions and policies: Help or hindrance to economic development? *The World Bank Economic Review, 6*(Suppl. 1), 117–144.

Garzón de la Roza, G. (2006). *Continued struggle for survival: How Plan Jefes y Jefas affected poor women's lives in Greater Buenos Aires, 2002–2005* (PhD thesis), Georgetown University Department of Government, Washington, DC.

Giavazzi, F., Jappelli, T., & Pagano, M. (2000). Searching for non-linear effects of fiscal policy: Evidence from industrial and developing countries. *European Economic Review, 44*(7), 1259–1289.

Gonzalez, L. (2016, May 12). South Africa to introduce new TB treatment. *Health24*. Accessible at: https://www.health24.com/Medical/Tuberculosis/TB-in-south-africa/south-africa-to-introduce-new-tb-treatment-20160329.

Gumede, W. (2005). *Thabo Mbeki and the battle for the soul of the ANC*. Cape Town: Zebra Press.

Hart, G. (2012). *Rethinking the South African crisis: Nationalism, populism, hegemony*. Durban: UKZN Press.

Head, T. (2019, February 12). Eskom: Experts reveal how much money load shedding is costing South Africa. *The South African*. Accessible at: https://www.thesouthafrican.com/eskom-how-much-money-does-load-shedding-cost-south-africa/.

Hello Doctor. (2017, April 18). *The cost of tuberculosis*. Accessible at: https://www.hellodoctor.co.za/the-cost-of-tuberculosis/.

IADRC. (2003, May 1). *South Africa: Apartheid debt and reparations*. International Apartheid Debt and Reparations Campaign. Accessible at: https://www.pambazuka.org/governance/south-africa-apartheid-debt-and-reparations.

Kanabus, A. (2018). *Information about tuberculosis, GHE*. Accessible at: https://www.tbfacts.org/tb-statistics-south-africa/.

Khera, R., & Nayak, N. (2009). Women workers and perceptions of the National Rural Employment Guarantee Act. *Economic and Political Weekly, 44*(43), 49–57.

Klein, N. (2012). *Real wage, labor productivity, and employment trends South Africa*. Washington, DC: IMF Africa Department.

Krohn, M. (1976, June). Inequality, unemployment and crime: A cross-national analysis. *The Sociological Quarterly, 17*(3), 303–313.

Laing, R. (2018, November 1). South African factories running at four-fifths capacity. *Business Day.* Accessible at: https://www.businesslive.co.za/bd/economy/2018-11-01-south-african-factories-running-at-four-fifths-capacity/.

Lawrence, A. (2019, February 26). Why a guaranteed job scheme in South Africa would pay for itself. *The Conversation.* Accessible at: https://theconversation.com/why-a-guaranteed-jobs-scheme-in-south-africa-would-pay-for-itself-112243.

Makou, G. (2018, February 7). Femicide in South Africa: 3 numbers about the murdering of women investigated. *Africa Check.* Accessible at: https://africacheck.org/reports/femicide-sa-3-numbers-murdering-women-investigated/.

Marais, H. (2018). The employment crisis, just transition, and the basic income grant. In V. Satgar (Ed.), *The climate crisis: South African and global democratic eco-socialist alternatives* (pp. 70–106). Johannesburg: Wits University Press.

Mendonça, M., Jacobs, D., & Sovacool, B. (2009). *Powering the Green Economy: The Feed-in Tariff Handbook.* London: Earthscan Limited.

Merten, M. (2019, March 1). State capture wipes out third of SA's R4.9-trillion GDP—Never mind lost trust, confidence, opportunity. *Daily Maverick.* Accessible at: https://www.dailymaverick.co.za/article/2019-03-01-state-capture-wipes-out-third-of-sas-r4-9-trillion-gdp-never-mind-lost-trust-confidence-opportunity/.

Mosler, W. (1997). Full employment and price stability. *Journal of Post Keynesian Economics, 20*(2), 167–182.

Nestle, U. (2014). Marktrealität von Bürgerenergie und mögliche Auswirkungen von regulatorischen Eingriffen. *Eine Studie für das Bündnis Bürgerenergie e.V. (BBEn) und dem Bund für Umwelt und Naturschutz Deutschland e.V. (BUND).* Accessible at: https://www.enklip.de/resources/Studie_Marktrealitaet+von+Buergerenergie_Leuphana_FINAL_23042014.pdf.

OEC. (2018). *South Africa – Country Profile.* Accessible at: https://atlas.media.mit.edu/en/profile/country/zaf/.

Okudoh, V., Trois, C., Workneh, T., & Schmidt, S. (2014). The potential of cassava biomass and applicable technologies for sustainable biogas production in South Africa: A review. *Renewable and Sustainable Energy Reviews, 39*, 1035–1052.

Omarjee, L. (2018, March 9). Culture of non-payment threatens stability of municipal finances—Treasury. *Fin24.* Accessible at: https://www.fin24.com/Economy/culture-of-non-payment-threatens-stability-of-municipal-finances-treasury-20180308.

One Million Climate Jobs. (n.d.). *One Million Climate Jobs: A just transition to a low carbon economy to combat unemployment and climate change.* https://womin.org.za/images/the-alternatives/fighting-destructive-extractivism/One%20Million%20Climate%20Jobs.pdf.

Patnaik, U. (2003). Global capitalism, deflation and agrarian crisis in developing countries. *Journal of Agrarian Change, 3*(1–2), 33–66.

Ramaphosa, C. (2018, October 4). *Address by President Cyril Ramaphosa at the opening of the Presidential Jobs Summit, Gallagher Estate, Johannesburg.* Accessible at: https://www.gov.za/speeches/president-cyril-ramaphosa-2019-state-nation-address-7-feb-2019-0000.

Rogerson, C. M. (2010). Local economic development in South Africa: Strategic challenges. *Development Southern Africa, 27*(4), 481–495.

Roopnarain, A., & Adeleke, R. (2017). Current status, hurdles and future prospects of biogas digestion technology in Africa. *Renewable and Sustainable Energy Reviews, 67,* 1162–1179.

Rose, P. (2013). Long-term sustainability in the management of acid mine drainage wastewaters—Development of the Rhodes BioSURE Process. *Water SA, 39*(5), 582–592.

Saba, A., Pather, R., & Macupe, B. (2018, March 16). Another child dies in a pit latrine. *Mail & Guardian.* Accessible at: https://mg.co.za/article/2018-03-16-00-another-child-dies-in-a-pit-latrine.

SALGA. (n.d.). *Local government energy efficiency and renewable energy strategy: Full strategy document.* Cape Town. Accessible at: http://sustainable.org.za/sustainable.org.za/uploads/files/file100.pdf.

SAMI. (2018). *GDP growth calculator.* South African Market Insights. Accessible at: https://www.southafricanmi.com/gdp-growth-estimator.html.

South Gauteng High Court. (2016). *Nkala and Others v Harmony Gold Mining Company Limited and Others.* Accessible at: http://www.saflii.org/za/cases/ZAGPJHC/2016/97.html.

StatsSA. (2017). *Non-financial census of municipalities report, 2017.* Accessible at: http://www.statssa.gov.za/?p=11199.

Stefan, A., & Paul, L. (2008). Does it pay to be green? A systematic overview. *Academy of Management Perspectives, 22*(4), 45–62.

Stockhammer, E. (2015). Rising inequality as a cause of the present crisis. *Cambridge Journal of Economics, 39*(3), 935–958.

Styan, J. (2015). *Blackout: The Eskom crisis.* Cape Town: Jonathan Ball.

Tcherneva, P. R., & Wray, L. R. (2005, December). *Gender and the job guarantee: The impact of Argentina's Jefes program on female heads of poor households* (Working Paper No. 50). Kansas City, MO: Center for Full Employment and Price Stability. Accessible at: http://www.cfeps.org/pubs/wp-pdf/WP50-Tcherneva-Wray.pdf.

Theron, A. (2018, February 5). Copper theft remains a serious concern—Eskom. *ESI Africa.* Accessible at: https://www.esi-africa.com/copper-theft-remains-a-serious-concern-eskom/.

Tshwane, T. (2018, October 4). Ramaphosa's jobs summit plan to create 275 000 new jobs annually. *Mail & Guardian.* Accessible at: https://mg.co.za/

article/2018-10-04-ramaphosas-jobs-summit-plan-to-create-275-000-new-jobs-annually.

Valodia, I., & Francis, D. (2016, November 28). How the search for a national minimum wage laid bare South Africa's faultlines. *The Conversation*. Accessible at: https://theconversation.com/how-the-search-for-a-national-minimum-wage-laid-bare-south-africas-faultlines-69382.

WaterAid. (2017). Out of Order The State of the World's Toilets 2017. Accessible at: http://www.wateraid.se/dokument/Out_of_Order_report_2017.pdf.

Western Cape Province. (2019). *Upgrade to a grid-tied solar PV system*. Accessible at: https://www.westerncape.gov.za/energy-security-game-changer/upgrade-grid-tied-solar-pv-system.

Wheels24. (2017, May 4). SA fuel price breakdown: Here's all you need to know. *Wheels24*. Accessible at: https://www.wheels24.co.za/AutoTrader/sa-petrol-price-heres-how-much-fuel-will-cost-you-20170405.

World Bank. (2017, March 23). *Tackling tuberculosis in Southern Africa's mineworkers with an innovative approach*. Washington, DC: World Bank Group. Accessible at: http://www.worldbank.org/en/news/feature/2017/03/23/tackling-tuberculosis-in-southern-africas-mineworkeres-with-an-innovative-approach.

References

Adler, E., Barnett, M., & Smith, S. (Eds.). (1998). *Security communities.* New York: Cambridge University Press.

AIDC. (2016). *One million climate jobs: Moving South Africa forward on a low-carbon, wage-led, and sustainable path.* Accessible at: http://aidc.org.za/download/climate-change/OMCJ-booklet-AIDC-electronic-version.pdf.

Alda, E., & Cuesta, J. (2011). A comprehensive estimation of costs of crime in South Africa and its implications for effective policy making. *Journal of International Development, 23*(7), 926–935.

Alfreds, D. (2018, April 13). Rural North West school gets green electricity. *News24.* Accessible at: https://www.news24.com/Green/News/rural-north-west-school-gets-green-electricity-20180413.

Amelang, S. (2016, June 29). Germany's energy transition revamp stirs controversy over speed, participation. *Clean Energy Wire.* Accessible at: https://www.cleanenergywire.org/dossiers/reform-renewable-energy-act.

ANC Presidency. (2011). Accessible at: http://www.thepresidency.gov.za/speeches/address-president-jacob-zuma-occasion-new-age-business-breakfast-cape-town-leg%2C-cape-town.

Arent, D., Arndt, C., Miller, M., & Zinaman, O. (Eds.). (2017). *The Political Economy of Clean Energy Transitions.* Oxford: Oxford University Press.

Ashley, B. (2018). Climate jobs at two minutes to midnight. In V. Satgar (Ed.), *The climate crisis: South African and global democratic eco-socialist alternatives* (pp. 272–292). Johannesburg: Wits University Press.

Ashman, S., Fine, B., & Newman, S. (2010). The crisis in South Africa: Neoliberalism, financialization and uneven and combined development. *Socialist Register, 47,* 174–195.

© The Editor(s) (if applicable) and The Author(s) 2020
A. Lawrence, *South Africa's Energy Transition*, Progressive Energy Policy, https://doi.org/10.1007/978-3-030-18903-7

Ashman, S., Fine, B., & Newman, S. (2011). Amnesty International? The nature, scale and impact of capital flight from South Africa. *Journal of Southern African Studies, 37*(1), 7–25. https://doi.org/10.1080/03057070.2011.555155.

Auty, R. (1993). *Sustaining development in mineral economies: The resource curse thesis.* New York: Oxford University Press.

Baker, L. (2015, August). The evolving role of finance in South Africa's renewable energy sector. *Geoforum.* https://doi.org/10.1016/j.geoforum.2015.06.017.

Barta, B. (2018, March 9). The contribution of pumped storage schemes to energy generation in South Africa. *Creamer Media's Engineering News.* Accessible at: https://www.ee.co.za/article/the-contribution-of-pumped-storage-schemes-to-energy-generation-in-south-africa.html.

Basson, A., & Du Toit, P. (2017). *Enemy of the people.* Cape Town: Jonathan Ball.

BBC. (2018a, August 14). South Africa to eradicate pit latrine toilets in schools. *BBC News.* Accessible at: https://www.bbc.co.uk/news/world-africa-45183593.

BBC. (2018b, July 6). South African boy of three drowns in Limpopo toilet. *BBC News.* Accessible at: https://www.bbc.co.uk/news/world-africa-44736772.

BBC. (2018c, September 18). South Africa crime: Can the country be compared to a 'war zone'? *BBC News.* Accessible at: https://www.bbc.co.uk/news/world-africa-45547975.

Beerten, J., Laes, G., Meskens, G., & D'haeseleer, W. (2009, December). Greenhouse gas emissions in the nuclear life cycle: A balanced appraisal. *Energy Policy, 37*(12), 5056–5068.

Bhorat, H., & Swilling, M. (2017). *Betrayal of the promise: How South Africa is being stolen* (State Capacity Research Project).

Bickerstaff, K., Walker, G., & Bulkeley, H. (Eds.). (2013). *Energy justice in a changing climate: Social equity and low-carbon energy.* New York: Zed Books.

Blecker, R. (2002). Kaleckian macro models for open economies. In J. Deprez & J. Harvey (Eds.), *Foundations of international economics: Post-Keynesian perspectives* (pp. 126–160). London: Routledge.

Bischof-Niemz, T. (2015, June). *How to stimulate the South African rooftop PV market without putting municipalities' financial stability at risk: A Net Feed-in Tariff proposal.* CSIR/02400/Renewable Energy/IR/2015/0002/A. Accessible at: http://www.wind-works.org/cms/fileadmin/user_upload/Files/Reports/South_Africa_NETFIT_concept_-_CSIR_-_8Jun2015.pdf.

Bloom, K. (2018, November 07). Interview: South Africa, climate change 'hot spot'. *Daily Maverick.* Accessible at: https://www.dailymaverick.co.za/article/2018-11-07-interview-south-africa-climate-change-hot-spot/.

Bond, P. (2000). *Elite transition: From apartheid to neoliberalism in South Africa.* London: Pluto Press.

Bond, P. (2019, January 1). South Africa suffers capitalist crisis Déjà Vu. *Monthly Review*. Accessible at: https://monthlyreview.org/2019/01/01/south-africa-suffers-capitalist-crisis-deja-vu/.

Bond, P., & Dugard, J. (2008). The case of Johannesburg water: What really happened at the pre-paid 'Parish pump'. *Law, Democracy & Development, 12*(1), 1–28.

Bond, P., & Ngwane, T. (2010). Community resistance to energy privatisation in South Africa. In K. Abramsky (Ed.), *Sparking a worldwide energy revolution: Social struggles in the transition to a post-petrol world* (Chapter 15, pp. 197–207). Oakland, CA: AK Press.

Borchert, L. (2015, March 10). Germany between citizens' energy and Nimbyism. *Clean Energy Wire*. Accessible at: https://www.cleanenergywire.org/dossiers/peoples-energiewende.

Bratsis, P. (2014). Political corruption in the age of transnational capitalism. *Historical Materialism, 22*(1), 105–128.

Braunstein, E., & Heintz, J. (2008). Gender bias and central bank policy: Employment and inflation reduction. *International Review of Applied Economics, 22*(2), 173–186.

Breytenbach, M. (2017, December 8). SAPVIA launches 'bold' 5-point plan for solar industrialization. *Creamer Media's Engineering News*. Accessible at: http://www.engineeringnews.co.za/article/sapvia-launches-bold-5-point-plan-for-solar-industrialisation-2017-12-08.

Brkic, B. (2010, April 7). High noon for Eskom's World Bank loan bid. *Daily Maverick*. Accessible at: https://www.dailymaverick.co.za/article/2010-04-07-high-noon-for-eskoms-world-bank-loan-bid/.

Budlender, D., Casale, D., & Valodia, I. (2010). Gender equality and taxation in South Africa. In C. Grown & I. Valodia (Eds.), *Taxation and gender equity: A comparative analysis of direct and indirect taxes in developing and developed countries* (pp. 206–232). New York: Routledge.

Burton, J., Lott, T., & Rennkamp, B. (2018). Sustaining carbon lock-in: Fossil fuel subsidies in South Africa. In J. Skovgaard & H. van Asselt (Eds.), *The politics of fossil fuel subsidies and their reform* (pp. 229–245). New York: Cambridge University Press.

Busby, C., & Mangano, J. (2017). There's a world going on underground—Infant mortality and fracking in Pennsylvania. *Journal of Environmental Protection, 8*, 381–393. https://doi.org/10.4236/jep.2017.84028.

BusinessTech. (2016, June 10). How much money violent crime costs South Africa every year. *BusinessTech*. Accessible at: https://businesstech.co.za/news/general/126263/how-much-money-violent-crime-costs-south-africa-every-year/.

BusinessTech. (2017, June 9). South Africa's shocking road death numbers at highest level in 10 years. *BusinessTech*. Accessible at: https://businesstech.co.

za/news/motoring/178275/south-africas-shocking-road-death-numbers-at-highest-level-in-10-years/.

Carbone, G., & Memoli, V. (2015). Does democratization foster state consolidation? Democratic rule, political order, and administrative capacity. *Governance, 28*(1), 5–24.

Carnie, T. (2015, July 16). At least one nuke power station for KZN. *IOL News*. Accessible at: http://www.iol.co.za/news/south-africa/kwazulu-natal/at-least-one-nuke-power-station-for-kzn-1886079.

Carr, M. (2018, November 30). Almost half coal power plants seen unprofitable to operate. *Bloomberg*. Accessible at: https://www.bloomberg.com/news/articles/2018-11-30/almost-half-of-coal-power-plants-seen-unprofitable-to-operate.

CDE. (2008, July). *South Africa's electricity crisis: How did we get here? And how do we put things right?* (CDE Roundtable No. 10). Johannesburg: Centre for Development and Enterprise.

Christie, R. (1984). *Electricity, industry and class in South Africa*. London: Palgrave Macmillan.

Clean Cooking Alliance. (2018). *South Africa*. Accessible at: http://cleancookingalliance.org/country-profiles/37-south-africa.html.

Climate Action Tracker. (2018). *South Africa*. Accessible at: https://climateactiontracker.org/countries/south-africa/.

Cloete, K. (2018, March 15). Our problem with the IPPs. *PoliticsWeb*. Accessible at: https://www.politicsweb.co.za/opinion/our-problem-with-the-ipps.

Coady, D., Parry, I., Sears, L., & Shang, B. (2017, March). How large are global fossil fuel subsidies? *World Development, 91*, 11–27. https://doi.org/10.1016/j.worlddev.2016.10.004.

Cottle, E. (2019). Competing Marxist Theories on the Temporal Aspects of Strike Waves: Silver's Product Cycle Theory and Mandel's Long Wave Theory. *Global Labour Journal, 10(1)*, 37–50.

Cotterill, J. (2018, June 14). Power outages hit South Africa as electricity monopoly cuts supply. *Financial Times*. Accessible at: https://www.ft.com/content/63a6bd76-6ff4-11e8-852d-d8b934ff5ffa.

Creamer, T. (2011, June 23). Fresh concern that SA is poised to abandon REFIT in favour of competitive bidding. *Creamer Media's Engineering News*. Accessible at: http://www.engineeringnews.co.za/article/fresh-concern-that-SA-is-poised-to-abandon-reFIT-in-favour-of-competitive-bidding-2011-06-23.

Creamer, T. (2013, May 16). Eskom awards 46 wind-turbine contract for Sere Wind Farm. *Creamer Media's Engineering News*. Accessible at: http://www.engineeringnews.co.za/article/eskom-awards-46-wind-turbine-contract-for-sere-wind-farm-2013-05-16/rep_id:4136.

Creamer, T. (2016, July 21). Eskom letter sends shock waves through private power sector. *Creamer Media's Engineering News*. Accessible at: http://www.

engineeringnews.co.za/article/eskom-letter-sends-shock-waves-through-private-power-sector-2016-07-21.

CSIR. (2015a, January 21). *2014 sees financial benefits of renewable energy exceed costs in South Africa.* Accessible at: https://www.csir.co.za/2014-sees-financial-benefits-renewable-energy-exceed-costs-south-africa.

CSIR. (2015b, August 19). *First half of 2015 sees financial benefits from renewable energy with huge cost savings.* Accessible at: https://www.csir.co.za/first-half-2015-sees-financial-benefits-renewable-energy-huge-cost-savings.

Davis, G. (1995). Learning to love the Dutch disease: Evidence from mineral economies. *World Development, 23*(10), 1765–1779.

DEA (Department of Environmental Affairs). (2014). *Greenhouse Gas Inventory for South Africa 2000–2010.* Pretoria: DEA.

Desai, A., Maharaj, B., & Bond, P. (2011). Introduction: Poverty eradication as holy grail. In B. Maharaj, A. Desai, & Bond, P. (Eds.), *Zuma's Own Goal: Losing South Africa's "War on Poverty"* (pp. 1–35). Trenton, NJ: Africa World Press.

Department of Energy (DoE). (2009). *Creating an enabling environment for distributed power generation in the South African electricity supply industry.* Pretoria: DoE.

Department of Energy (DoE). (2012, September 27). Postponement of 3rd bid submission date for the REIPPP. *EE Publishers.* Accessible at: http://www.ee.co.za/article/deptenergy-359-10-postponement-of-3rd-bidsubmission-date-for-the-reipppp.html.

Department of Energy (DoE). (2016). *Integrated energy plan annexure B: Macroeconomic assumptions.* Accessible at: http://www.energy.gov.za/files/IEP/2016/IEP-AnnexureB-Macroeconomic-Assumptions.pdf.

Department of Environmental Affairs (DEA). (2014). *Greenhouse gas inventory for South Africa 2000–2010.* Pretoria: DEA.

Department of Environmental Affairs (DEA). (2018, March). *South Africa's third national communication under the United Nations framework convention on climate change.* Pretoria: DEA. Accessible at: https://unfccc.int/sites/default/files/resource/South%20African%20TNC%20Report%20%20to%20the%20UNFCCC_31%20Aug.pdf.

Department of Mineral Resources (DMR). (2014). South Africa's coal industry—Overview, 2014 (Report R111). Accessible at: http://www.dmr.gov.za/LinkClick.aspx?fileticket=uePTS1drX6A%3D&portalid=0.

Department of Water Affairs and Forestry (DWAF). (2009). *Water resources analysis, Vaal River system: Large bulk water supply reconciliation strategy* (DWAF Report No. PRSA C000/00/4406/05). Pretoria: DWAF.

Deign, J. (2018, July). *South Africa Open Again for Renewables After Auction Turmoil, Greentech Media.* Accessible at: https://www.greentechmedia.com/articles/read/south-africa-open-again-for-renewables-after-auction-turmoil.

Deininger, K., & Liu, Y. (2013). *Welfare and poverty impacts of India's national rural employment guarantee scheme: Evidence from Andhra Pradesh.* Washington, DC: The World Bank.

Deutsch, K. (1961). Security communities. In J. Roseau (Ed.), *International politics and foreign policy: A reader in research and theory.* Glencoe, NY: Free Press.

Dhansay, T., Musekiwa, C., Ntholi, T., Chevallier, L., Cole, D., & de Wit, M. J. (2017). South Africa's geothermal energy hotspots inferred from subsurface temperature and geology. *South African Journal of Science.* http://doi.org/10.17159/sajs.2017/20170092.

DME. (1998). *White paper on the energy policy of the Republic of South Africa.* Department of Minerals and Energy, Government of South Africa. Accessible at: http://www.energy.gov.za/files/policies/whitepaper_energypolicy_1998.pdf.

DME. (2003). *White paper on renewable energy.* Department of Minerals and Energy, Government of South Africa. Accessible at: https://unfccc.int/files/meetings/seminar/application/pdf/sem_sup1_south_africa.pdf.

Department of Energy (DoE). (2003). *South African Department of Energy, White Paper on Renewable Energy.* Accessible at: https://www.energy.gov.za/files/policies/whitepaper_renewables_2003.pdf.

Department of Energy (DoE). (2011a). *South African Department of Energy, Integrated Resource Plan for Electricity, 2010–2030.*

Department of Energy (DoE). (2011b). *Electricity Regulations on New Generation Capacity.* Accessible at: https://www.energy.gov.za/files/policies/Electricity%20Regulations%20on%20New%20Generation%20Capacity%201-34262%204-5.pdf.

Department of Energy (DoE). (2014). *The REIPP Procurement Programme Part B: Qualification criteria.* Republic of South Africa.

Donnelly, L. (2009, November 6). Eskom upheaval: The case against Maroga. *Mail & Guardian.* Accessible at: https://mg.co.za/article/2009-11-06-eskom-upheaval-the-case-against-maroga.

DPLG (Department of Provincial and Local Government). (2006). *Stimulating and Developing Sustainable Local Economies: National Framework for Local Economic Development (LED) in South Africa.* Pretoria: DPLG.

DPRU. (2018, June 2). *Monitoring the performance of the South African labour market: An overview of the South African labour market for the year ending 2017 quarter 2.* Development Policy Research Unit, University of Cape Town. Accessible at: http://www.dpru.uct.ac.za/sites/default/files/image_tool/images/36/Publications/Other/2018-06-13%20Factsheet%2021%20-%20Year%20ended%202017Q2.pdf.

DWA. (2012). *Full technical report on the implication of climate change impacts on water resources planning in South Africa.* Pretoria: Department of Water Affairs.

Eberhard, A. (2005). From state to market and back again: South Africa's power sector reforms. *Economic and Political Weekly, 40*(50), 5309–5317.

Ebarhard, A. (2007). The political economy of power sector reform in South Africa. In D. Victor & T. Heller (Eds.), *The Political Economy of Power Sector Reform* (pp. 215–253). New York: Cambridge University Press.

Eberhard, A. (2011). *The future of South African coal: Market, investment, and policy challenges* (Working Paper No. 100). Stanford: Stanford University Program on Energy and Sustainable Development.

Eberhard, A., Kolker, J., & Leigland, J. (2014).*South Africa's renewable energy IPP procurement program: Success factors and lessons*. Accessible at: https://www.gsb.uct.ac.za/files/ppiafreport.pdf.

Eberhard, A., & Naude, R. (2016). The South African renewable energy independent power producer procurement programme: A review and lessons learned. *Journal of Energy in Southern Africa, 27*(4). Accessible at: http://www.scielo.org.za/scielo.php?script=sci_arttext&pid=S1021-447X2016000400001.

Economist. (1999, April 22). South Africa's debt: Unforgivable. *Economist*. Accessible at: https://www.economist.com/finance-and-economics/1999/04/22/unforgivable.

Edigheji, O. (Ed.). (2010). *Constructing a developmental state in South Africa.* Cape Town: HSRC Press.

EIA. (2016). *Global energy intensity continues to decline.* Washington, DC: US Energy Information Administration. Accessible at: https://www.eia.gov/todayinenergy/detail.php?id=27032.

Ellis, S. (2013). *External mission: The ANC in exile, 1960–1990.* Oxford: Oxford University Press.

EMG. (n.d.). *Water management devices: Facts and perspectives.* Cape Town: Environmental Monitoring Group. Accessible at: http://www.emg.org.za/images/downloads/water_cl_ch/FactSheetWMD.pdf.

ENCA. (2013, March 22). Eskom-BHP contract a 'scandal'. *ENCA*. Accessible at: https://www.enca.com/money/eskom-bhp-contract-scandal.

ENCA. (2015, June 5). How does biogas technology work? *ENCA*. Accessible at: https://www.enca.com/south-africa/gallery-how-does-biogas-technology-work.

Energati. (2016, January 19). South Africa's solar market needs innovative financing. *Energati*. Accessible at: http://www.engerati.com/article/south-africa%E2%80%99s-solar-market-needs-innovative-financing.

Energy News. (2018, May 30). Report: SA doesn't need R19bn coal IPP projects. *Energy News*. Accessible at: https://www.bizcommunity.com/Article/196/693/177655.html.

Epstein, M. J., & Buhovac, A. R. (2014). *Making sustainability work: Best practices in managing and measuring corporate social, environmental, and economic impacts.* San Francisco: Berrett-Koehler Publishers.

EScience Associates, UrbanECON, & Ahlfeldt, C. (2013). *The localisation potential of photovoltaics (PV) and a strategy to support large scale roll-out in*

South Africa (Report prepared for SAPVIA, DTi, WWF, South Africa). Accessible at: http://www.sapvia.co.za/wp-content/uploads/2013/04/PV-Localisation_Draft-FinalReport-v1.2.pdf.

Eskom. (2005). *Annual Report, 2004-2005*. Accessible at: http://www.eskom.co.za/sites/heritage/Annual%20Reports/2004-2005%20Annual%20Report.pdf.

Esswein, E., Breitenstein, M., Snawder, J., Kiefer, M., & Sieber, W. (2013). Occupational exposures to respirable crystalline silica during hydraulic fracturing. *Journal of Occupational and Environmental Hygiene, 10*(7), 347–356.

Etheridge, J. (2017, March 30). Government gives green light for shale gas fracking in Karoo. *News24*. Accessible at: https://www.news24.com/SouthAfrica/News/govt-gives-green-light-for-shale-gas-fracking-in-karoo-20170330?fbclid=IwAR0L1O_QIpA4KUCRWxH53EqZiFoxmgvN43RtEkHVy-qrUfwahtX8yhB1M18.

Evans, P. (2010). Constructing the 21st century developmental state: Potentialities and pitfalls. In O. Edighej (Ed.), *Constructing a democratic developmental state in South Africa: Potentials and challenges* (pp. 37–59). Cape Town: HSRC Press.

Fajnzylber, D., Lederman, D., & Loayza, N. (2002, April). Inequality and violent crime. *Journal of Law and Economics, 45*, 1–40.

Feinstein, A. (2007). *After the party: A personal and political journey inside the ANC*. Cape Town: Jonathan Ball.

Fig, D. (2005). *Uranium road: Questioning South Africa's nuclear direction*. Cape Town: Jacana Media.

Fig, D. (2018). Capital, climate and the politics of nuclear procurement in South Africa. In V. Satgar (Ed.), *The climate crisis: South African and global democratic eco-socialist alternatives* (pp. 252–271). Johannesburg: Wits University Press.

Fiil-Flynn, M., & Soweto Electricity Crisis Committee. (2001). *Electricity crisis in Soweto* (Municipal Services Project Occasional Papers No. 4), 1.

Fine, B., & Rustomjee, Z. Z. R. (1996). *The political economy of South Africa: From Minerals Energy Complex to Industrialisation*. Boulder, CO: Westview.

Fin24. (2014, November 23). Eskom issues power emergency. *News24*. Accessible at: https://www.fin24.com/Economy/Eskom-issues-power-emergency-20141123.

Foster, N., Vassall, A., Cleary, S., Cunnama, L., Churchyard, G., & Sinanovic, E. (2015). The economic burden of TB diagnosis and treatment in South Africa. *Social Science & Medicine, 130*, 42–50.

Freeman, R. B. (1992). Labor market institutions and policies: Help or hindrance to economic development? *The World Bank Economic Review, 6*(Suppl. 1), 117–144.

Garzón de la Roza, G. (2006). *Continued struggle for survival: How Plan Jefes y Jefas affected poor women's lives in Greater Buenos Aires, 2002–2005* (PhD thesis), Georgetown University Department of Government, Washington, DC.

Geels, F. (2002). Technological transitions as evolutionary reconfiguration processes: A multi-level perspective and case study. *Research Policy, 31*(8–9), 1257–1274.

Geels, F. (2018). Disruption and low-carbon system transformation: Progress and new challenges in socio-technical transitions research and the multi-level perspective. *Energy Research & Social Science, 37,* 224–231.

Geels, F. W. (2005). *Technological Transitions and System Innovations: A Co-Evolutionary and Socio-Technical analysis.* Edward Elgar Publishing.

Geels, F. W. (2010). Ontologies, socio-technical transitions (to sustainability), and the multi-level perspective. *Research policy, 39*(4), 495–510.

Geels, F. W. (2014). Regime resistance against low-carbon transitions: Introducing politics and power into the multi-level perspective. *Theory, Culture & Society, 31*(5), 21–40.

Giavazzi, F., Jappelli, T., & Pagano, M. (2000). Searching for non-linear effects of fiscal policy: Evidence from industrial and developing countries. *European Economic Review, 44*(7), 1259–1289.

Gibbs, A., & Lloyd, T. (2018, October 25). *The myth of "clean coal": Why coal can only ever be dirty.* EE Publishers. Accessible at: http://www.ee. co.za/article/the-myth-of-clean-coal-why-coal-can-only-ever-be-dirty.html#. W9KzDxi6I0M.

GHS-SA. (2016). General household survey (GHS) 2016 report. *StatsSA.* Accessible at: http://www.statssa.gov.za/publications/P0318/P03182017.pdf.

Gleason, D. (2013, July 10). Medupi farce may cost trillion-plus. *Business Day Live.* Accessible at: https://www.businesslive.co.za/bd/opinion/columnists/2013-07-10-medupi-farce-may-cost-trillion-plus/.

Gonzalez, L. (2016, May 12). South Africa to introduce new TB treatment. *Health24.* Accessible at: https://www.health24.com/Medical/Tuberculosis/TB-in-south-africa/south-africa-to-introduce-new-tb-treatment-20160329.

Gosling, M. (2017, February 25). Russian nuclear deal places massive liability on South Africans. *News24.* Accessible at: https://www.news24.com/SouthAfrica/News/russian-nuclear-deal-places-massive-liability-on-south-africans-20170225.

Gramsci, A. (1992). *Prison notebooks* (J. Buttigieg, Trans., Vol. 1). New York: Columbia University Press.

Grubler, A. (2012). Energy transitions research insights and cautionary tales. *Energy Policy, 50,* 8–16.

Gumede, W. (2005). *Thabo Mbeki and the battle for the soul of the ANC.* Cape Town: Zebra Press.

Habib, A. (2013). *South Africa's Suspended Revolution: Hopes and Prospects.* Johannesburg: Wits University Press.

Harrington, R. (2016, October 9). Where Hillary Clinton stands on climate change. *Business Insider.* Accessible at: https://www.businessinsider.com/

hillary-clinton-environment-climate-change-platforms-policies-plans-2016-10?
r=US&IR=T.

Hart, G. (2012). *Rethinking the South African crisis: Nationalism, populism, hegemony*. Durban: UKZN Press.

Hartnady, C. (2010). South Africa's diminishing coal reserves. *South African Journal of Science, 106*(9–10). Accessible at: http://www.sajs.co.za.

Head, T. (2019, February 12). Eskom: Experts reveal how much money load shedding is costing South Africa. *The South African*. Accessible at: https://www.thesouthafrican.com/eskom-how-much-money-does-load-shedding-cost-south-africa/.

Heintz, J. (2010). The social structure of accumulation in South Africa. In T. McDonald, M. Reich, & D. Kotz (Eds.), *Contemporary capitalism and its crises* (pp. 267–285). New York: Cambridge University Press.

Hello Doctor. (2017, April 18). *The cost of tuberculosis*. Accessible at: https://www.hellodoctor.co.za/the-cost-of-tuberculosis/.

Hofstatter, S. (2018). *Licence to loot: How the plunder of Eskom and other parastatals almost sank South Africa*. Cape Town: Penguin Random House South Africa.

Holland, M. (2017, March 31). Health impacts of coal fired power plants in South Africa. *Groundwork South Africa/ Health Care Without Harm*. Accessible at: https://cer.org.za/wp-content/uploads/2017/04/Annexure-Health-impacts-of-coal-fired-generation-in-South-Africa-310317.pdf.

Howarth, R. W., Ingraffea, A., & Engelder, T. (2011). Natural gas: Should fracking stop? *Nature, 477*(7364), 271.

Humphreys, M., Sachs, J. D., & Stiglitz, J. E. (Eds.). (2007). *Escaping the resource curse*. New York: Columbia University Press.

Huntington, S. P. (1968). *Political order in changing societies*. New Haven: Yale University Press.

IADRC. (2003, May 1). *South Africa: Apartheid debt and reparations*. International Apartheid Debt and Reparations Campaign. Accessible at: https://www.pambazuka.org/governance/south-africa-apartheid-debt-and-reparations.

IEA. (2006). *World energy outlook*. Accessible at: https://www.iea.org/publications/freepublications/publication/weo2006.pdf.

IEA. (2017). *World energy investment 2017*. Accessible at: https://www.iea.org/publications/wei2017/.

InfluenceMap. (2018, December). Who Owns the World's Fossil Fuels? A forensic look at the operators and shareholders of fossil fuel companies. *InfluenceMap*. Accessible at: https://influencemap.org/finance-map.

IPCC. (2007). Climate change 2007: Impacts, adaptation and vulnerability. *Working Group II Contribution to the Fourth Assessment Report of the Intergovernmental Panel on Climate Change*. Accessible at: https://www.ipcc.ch/site/assets/uploads/2018/03/ar4_wg2_full_report.pdf.

IPCC. (2018). *Intergovernmental Panel on Climate Change interim report, 2018.* Accessible at: http://www.ipcc.ch/report/sr15/.

Ireland, G., & Burton, J. (2018). *An assessment of new coal plants in South Africa's electricity future: The cost, emissions, and supply security implications of the coal IPP programme.* Energy Research Centre, University of Cape Town, Cape Town, South Africa. Accessible at: https://cer.org.za/wp-content/uploads/2018/05/ERC-Coal-IPP-Study-Report-Finalv2-290518.pdf.

IRENA. (2018). *Renewable power generation costs in 2017.* Abu Dhabi: International Renewable Energy Agency. Accessible at: https://www.irena.org/-/media/Files/IRENA/Agency/Publication/2018/Jan/IRENA_2017_Power_Costs_2018.pdf.

IRP. (2010, May 6). *Integrated Resource Plan, 2010–2030* (Regulation Gazette No. 9531 GG No. 34263).

IRP. (2017). *South Africa's Integrated Resource Plan.* Accessible at: http://www.ee.co.za/wp-content/uploads/2017/12/Eskom-IRP-2017-study-report-for-DoE-November-2017.pdf.

IRP. (2018). *South Africa's Integrated Resource Plan.* Accessible at: http://www.energy.gov.za/IRP/irp-update-draft-report2018/IRP-Update-2018-Draft-for-Comments.pdf.

Jacobson, M. Z., Delucchi, M. A., Bauer, Z. A., Goodman, S. C., Chapman, W. E., Cameron, M. A., et al. (2017). 100% clean and renewable wind, water, and sunlight all-sector energy roadmaps for 139 countries of the world. *Joule, 1*(1), 108–121.

Jaglin, S., & Dubresson, A. (2016). *ESKOM: Electricity and Technopolitics in South Africa.* Cape Town: Juta and Company (Pty) Ltd.

Jiborn, M., Kander, A., Kulionis, V., Nielsen, H., & Moran, D. (2018). Decoupling or delusion? Measuring emissions displacement in foreign trade. *Global Environmental Change, 49,* 27–34. https://doi.org/10.1016/j.gloenvcha.2017.12.006.

Johnston, J., Lim, E., & Roh, H. (2019, March 20). Impact of upstream oil extraction and environmental public health: A review of the evidence. *Science of the Total Environment, 657,* 187–199.

Kalecki, M. (1971). Class struggle and the distribution of national income. *Kyklos, 24*(1), 1–9.

Kambule, N., Yessoufou, K., Nwulu, N., & Mbohwa, C. (2019). Exploring the driving factors of prepaid electricity meter rejection in the largest township of South Africa. *Energy Policy, 124,* 199–205.

Kanabus, A. (2018). *Information about tuberculosis, GHE.* Accessible at: https://www.tbfacts.org/tb-statistics-south-africa/.

Kasrils, R. (2018). *Armed and dangerous: From undercover struggle to freedom* (4th ed.). Johannesburg: Jacana Media.

Khera, R., & Nayak, N. (2009). Women workers and perceptions of the National Rural Employment Guarantee Act. *Economic and Political Weekly, 44*(43), 49–57.

Khumalo, S. (2018, April 4). Jeff Radebe signs R56bn contract with renewable power producers. *Fin24.* Accessible at: https://www.fin24.com/Economy/Eskom/jeff-radebe-signs-long-delayed-renewable-power-deals-20180404.

Kings, S. (2018, August 27). South Africa has a new energy plan. *Mail & Guardian.* Accessible at: https://mg.co.za/article/2018-08-27-south-africa-has-a-new-energy-plan.

Klein, N. (2012). *Real wage, labor productivity, and employment trends South Africa.* Washington, DC: IMF Africa Department.

Kohli, A. (2006). *State and development.* Cheltenham: Edward Elgar.

Krohn, M. (1976, June). Inequality, unemployment and crime: A cross-national analysis. *The Sociological Quarterly, 17*(3), 303–313.

Kuzemko, C., Lawrence, A., & Watson, M. (2019). The IPE of energy: Specificities, interactions, and contestations. *Review of International Political Economy, 26,* 80–103.

Lachapelle, E., MacNeil, R., & Paterson, M. (2017). The political economy of decarbonisation: From green energy 'race' to green 'division of labour'. *New Political Economy, 22*(3), 311–327.

Laing, R. (2018, November 1). South African factories running at four-fifths capacity. *Business Day.* Accessible at: https://www.businesslive.co.za/bd/economy/2018-11-01-south-african-factories-running-at-four-fifths-capacity/.

Lawrence, A. (2013). Neoliberalism, mineral resource governance and developmental states: South Africa in comparative perspective. In F. Bourgouin & J. Nem Singh (Eds.), *The political economy of extraction* (pp. 40–60). New York: Palgrave.

Lawrence, A. (2014). *Employer and worker and collective action: A comparative study of Germany, South Africa, and the United States.* New York: Cambridge University Press.

Lawrence, A. (2018). How can energy and climate justice claims be reconciled? In A. Goldthau, M. Keating, & C. Kuzemko (Eds.), *Handbook on the IPE of energy and resources* (pp. 227–237). Cheltenham: Edward Elgar.

Lawrence, A. (2019, February 26). Why a guaranteed job scheme in South Africa would pay for itself. *The Conversation.* Accessible at: https://theconversation.com/why-a-guaranteed-jobs-scheme-in-south-africa-would-pay-for-itself-112243.

Lazard. (2018). *Levelized cost of energy data.* Accessible at: https://www.lazard.com/media/450784/lazards-levelized-cost-of-energy-version-120-vfinal.pdf.

Leger, J. P. (1991). Trends and causes of fatalities in South African mines. *Safety Science, 14*(3–4), 169–185.

Le Roux, M. (2006, March 30). Mbeki: There is no electricity crisis. *Mail & Guardian*. Accessible at: https://mg.co.za/article/2006-03-30-mbeki-there-is-no-electricity-crisis.

Liefferink, M. (2016, November). *Rehabilitation of mine contaminated wetlands, eco-systems and receptor dams: A just transition to a low carbon economy to combat unemployment and climate change*. Cape Town: AIDC.

Lippit, V. (2010). Social structure of accumulation theory. In T. McDonough, M. Reich, & D. Kotz (Eds.), *Contemporary capitalism and its crises: Social structure of accumulation theory for the twenty first century* (pp. 45–71). New York: Cambridge University Press.

Llavador, H., Roemer, J. E., & Silvestre, J. (2015). *Sustainability for a warming planet*. Cambridge, MA: Harvard University Press.

Lodge, T. (2014, January 1). Neo-patrimonial politics in the ANC. *African Affairs, 113*(450), 1–23. https://doi.org/10.1093/afraf/adt069.

Loorbach, D. (2007). *Transition management: New mode of governance for sustainable development*. Utrecht, The Netherlands: International Books.

Mafeje, A. (1978). *Science, ideology, and development: Three essays on development theory*. Uppsala: Scandinavian Institute of Development Studies.

Magnani, M. B., Blanpied, M. L., DeShon, H. R., & Hornbach, M. J. (2017). Discriminating between natural versus induced seismicity from long-term deformation history of intraplate faults. *Science Advances, 3*(11), e1701593.

Mail & Guardian. (2010, April 3). Zille says ANC stands to make R1bn from Medupi. *Mail & Guardian*. Accessible at: https://mg.co.za/article/2010-04-03-zille-says-anc-stands-to-make-r1bn-from-medupi.

Makou, G. (2018, February 7). Femicide in South Africa: 3 numbers about the murdering of women investigated. *Africa Check*. Accessible at: https://africacheck.org/reports/femicide-sa-3-numbers-murdering-women-investigated/.

Malala, J. (2015). *We have now begun our descent: How to stop South Africa losing its way*. Cape Town: Jonathan Ball.

Manuel, T. (2011). Our diagnosis of SA's problems. *PoliticsWeb*. Accessible at: http://www.politicsweb.co.za/documents/our-diagnosis-of-sas-problems--trevor-manuel.

Marais, H. (2018). The employment crisis, just transition, and the basic income grant. In V. Satgar (Ed.), *The climate crisis: South African and global democratic eco-socialist alternatives* (pp. 70–106). Johannesburg: Wits University Press.

Marais, H. (2011). *South Africa pushed to the limit: The political economy of change*. London: Zed Books.

Marinovich, G. (2016). *Murder at small Koppie: The real story of the Marikana massacre*. Johannesburg: Penguin.

Martinez, D. M., & Ebenhack, B. W. (2008). Understanding the role of energy consumption in human development through the use of saturation phenomena. *Energy Policy, 36*(4), 1430–1435.

McDaid, L. (2014). *Renewable Energy Independent Power Producer Procurement Programme Review 2014.* Cape Town: Electricity Governance Initiative of South Africa.

McDaid, L. (2016). *Renewable energy independent power producer procurement programme review 2016: A critique of process of implementation of socio-economic benefits including job creation.* Cape Town: AIDC.

McKay, D. (2018, November 29). Exxaro was "frustrated" with Eskom coal procurement, but signs of hope detected. *Miningmx.* Accessible at: https://www.miningmx.com/news/energy/35291-exxaro-was-frustrated-with-eskom-coal-procurement-but-signs-of-hope-detected/.

McKinsey & Co. (2015).*South Africa's big five: Bold priorities for inclusive growth.* McKinsey Global Institute. Accessible at: https://studyres.com/doc/15857434/south-africa-s-big-five.

Mearsheimer, J. (1994). The false promise of international institutions. *International Security, 19*(3), 5–49.

Meier, P., Vagliasindi, M., & Mudassar, I. (2015). *The design and sustainability of renewable energy incentives: An economic analysis* (pp. 155–169). Washington, DC: World Bank Group.

Mendonça, M., Jacobs, D., & Sovacool, B. (2009). *Powering the Green Economy: The Feed-in Tariff Handbook.* London: Earthscan Limited.

Merten, M. (2019, March 1). State capture wipes out third of SA's R4.9-trillion GDP—Never mind lost trust, confidence, opportunity. *Daily Maverick.* Accessible at: https://www.dailymaverick.co.za/article/2019-03-01-state-capture-wipes-out-third-of-sas-r4-9-trillion-gdp-never-mind-lost-trust-confidence-opportunity/.

Meth, O. (2018). New satellite data reveals the world's largest air pollution hotspot is Mpumalanga – South Africa. *Greenpeace Africa.* Accessible at: https://www.greenpeace.org/africa/en/press/4202/new-satellite-data-reveals-the-worlds-largest-air-pollution-hotspot-is-mpumalanga-south-africa/.

Midgley, G., Chapman, R., Mukheibir, P., Tadross, M., Hewitson, B., Wand, S., et al. (2007). *Impacts, vulnerability and adaptation in key South African sectors: An input into the long-term mitigation scenarios process.* Cape Town: Energy Research Centre, University of Cape Town.

Mining Review. (2018, February 9). The future of "dirty" coal in South Africa vs reliable renewable energy sources. *Mining Review.* Accessible at: https://www.miningreview.com/future-dirty-coal-south-africa-reliable-renewable-energy-sources/.

Mohamed, S. (2010). The state of the South African economy. In J. Daniel, P. Naidoo, D. Pillay, & R. Southall (Eds.), *New South African review 1: Development or decline?* (pp. 39–65). Johannesburg: Wits University Press.

Montel. (2018). *South Africa Q1 thermal coal exports rise 6%.* Accessible at: https://www.montel.no/fr/story/south-africa-q1-thermal-coal-exports-rise-6/897919.

Montmasson-Clair, G., & Das Nair, R. (2017). South Africa's renewable energy experience: Inclusive growth lessons. In J. Klaaren, S. Roberts, & I. Valodia (Eds.), *Competition Law and Economic Regulation in Southern Africa: Addressing Market Power in Southern Africa* (pp. 97–119). Johannesburg: Wits University Press.

Montmasson-Clair, G., Moilwa, K., & Ryan, G. (2014). *Regulatory Entities Capacity Building Project Review of Regulators Orientation and Performance: Review of Renewable Energy Regulation.* Johannesburg and Pretoria: University of Johannesburg and Trade and Industrial Policy Strategies.

Monyei, C. G., Sovacool, B. K., Brown, M. A., Jenkins, K. E., Viriri, S., & Li, Y. (2019, January–February). Justice, poverty, and electricity decarbonization. *Electricity Journal, 32*(1), 47–51.

Mooney, C. (2011). The truth about fracking. *Scientific American, 305*(5), 80–85.

Mosler, W. (1997). Full employment and price stability. *Journal of Post Keynesian Economics, 20*(2), 167–182.

MPE. (2000). *An accelerated agenda toward the restructuring of state owned enterprises: Policy framework.* Pretoria: Ministry of Public Enterprises. Accessible at: https://www.gov.za/sites/default/files/gcis_document/201409/acceleratedagendarestructuringsoe0.pdf.

Mudavanhu, S., & Rudin, J. (2018, April 04). Renewable energy claims and counterclaims. *Daily Maverick.* Accessible at: https://www.dailymaverick.co.za/article/2018-04-04-op-ed-renewable-energy-claims-and-counterclaims/.

Mudd, G., & Diesendorf, M. (2008). Sustainability of uranium mining and milling: Toward quantifying resources and eco-efficiency. *Environmental Science & Technology, 42*(7), 2624–2629.

Muller, M. (2013). The regulation of network infrastructure beyond the Washington consensus. *Development Southern Africa, 30*(4–5), 674–686.

Mundy, S. (2019, January 1). India's renewable rush puts coal on the back burner. *Financial Times.* Accessible at: https://www.ft.com/content/b8d24c94-fde7-11e8-aebf-99e208d3e521.

Nace, T. (2018, October). A coal phase-out pathway for 1.5°C: Modeling a coal power phase-out pathway for 2018–2050 at the individual plant level in support of the ipcc 1.5°C findings on coal. *CoalSwarm.* Accessible at: https://storage.googleapis.com/planet4-international-stateless/2018/10/7df76ee5-coalpathway-final.pdf.

Nakumuryango, A., & Inglesi-Lotz, R. (2016). South Africa's performance on renewable energy and its relative position against the OECD countries and the rest of Africa. *Renewable and Sustainable Energy Reviews, 56,* 999–1007.

Nestle, U. (2014). Marktrealität von Bürgerenergie und mögliche Auswirkungen von regulatorischen Eingriffen. *Eine Studie für das Bündnis Bürgerenergie e.V. (BBEn) und dem Bund für Umwelt und Naturschutz Deutschland e.V. (BUND)*. Accessible at: https://www.enklip.de/resources/Studie_Marktrealitaet+von+Buergerenergie_Leuphana_FINAL_23042014.pdf.

Nathan, L. (2006). Domestic instability and security communities. *European Journal of International Relations, 12*(2), 275–299.

Newell, P. (2018). Trasformismo or transformation? The global political economy of energy transitions. *Review of International Political Economy*, 1–24. https://doi.org/10.1080/09692290.2018.1511448.

Newell, P., & Bumpus, A. (2012). The global political ecology of the CDM. *Global Environmental Politics, 12*, 49–67.

Newell, P., & Mulvaney, D. (2013). The political economy of the 'just transition'. *The Geographical Journal*. https://doi.org/10.1111/geoj.12008.

Nkosi, N., & Dikgang, J. (2018, January). *South African attitudes about nuclear power: The case of the nuclear energy expansion* (ERSA Working Paper No. 726). Pretoria. Accessible at: https://econrsa.org/2017/wp-content/uploads/working_paper_726.pdf.

NPC. (2012). *National Planning Commission, National Development Plan 2030: Our future—Make it work*. Pretoria: Government Printers.

Nxumalo, M. (2018, December 6). Environmental lobby prepares legal fight over offshore drilling. *Daily News*. Accessible at: https://www.iol.co.za/dailynews/environmental-lobby-prepares-legal-fight-over-offshore-drilling-18406392.

Ochieng, G., & Nkwonta, O. (2010, November 18). Impacts of mining on water resources in South Africa: A review. *Scientific Research and Essays, 5*(22), 3351–3357.

OEC. (2018). *South Africa—Country profile*. Accessible at: https://atlas.media.mit.edu/en/profile/country/zaf/.

Okudoh, V., Trois, C., Workneh, T., & Schmidt, S. (2014). The potential of cassava biomass and applicable technologies for sustainable biogas production in South Africa: A review. *Renewable and Sustainable Energy Reviews, 39*, 1035–1052.

Omarjee, L. (2018, March 9). Culture of non-payment threatens stability of municipal finances—Treasury. *Fin24*. Accessible at: https://www.fin24.com/Economy/culture-of-non-payment-threatens-stability-of-municipal-finances-treasury-20180308.

One Million Climate Jobs. (n.d.). *One Million Climate Jobs: A just transition to a low carbon economy to combat unemployment and climate change*. https://womin.org.za/images/the-alternatives/fighting-destructive-extractivism/One%20Million%20Climate%20Jobs.pdf.

Padayachee, V. (1991). The politics of South Africa's international financial relations, 1970–1990. In S. Gelb (Ed.), *South Africa's economic crisis*. Cape Town: David Philip.

Patnaik, U. (2003). Global capitalism, deflation and agrarian crisis in developing countries. *Journal of Agrarian Change, 3*(1–2), 33–66.

Pauw, J. (2017). *The president's keepers.* Cape Town: Tafelberg.

PFMA. (1999). *Public Finance Management Act of South Africa.* Accessible at: http://saflii.austlii.edu.au/za/legis/hist_act/pfma1o1999225/pfma1o1999a2s2013356.html.

Pieterse, C. (2018, August 21). Oil threat to KZN coast. *The Witness.*

PMG. (2001, June 11). *Eskom Conversion Bill: Cosatu input; Department briefing on additional amendments.* Accessible at: https://pmg.org.za/committee-meeting/596/.

Pollin, R. (2015). *Greening the global economy.* Cambridge: MIT Press.

Rademeyer, J. (2013). Claim that 94% in SA have access to safe drinking water…Doesn't hold water. *AfricaCheck.* Accessible at: https://africacheck.org/reports/claim-that-94-of-south-aclaim-that-94-in-sa-have-access-to-safe-drinking-water-doesnt-hold-water/.

Ramaphosa, C. (2018, October 4). *Address by President Cyril Ramaphosa at the opening of the Presidential Jobs Summit, Gallagher Estate, Johannesburg.* Accessible at: https://www.gov.za/speeches/president-cyril-ramaphosa-2019-state-nation-address-7-feb-2019-0000.

Ravallion, M. (2007). Inequality is bad for the poor. In S. Jenkins & J. Micklewright (Eds.), *Inequality and poverty re-examined.* Oxford: Oxford University Press.

Rizvi, F. (2019, February 11). South Africa ready to discuss nuclear energy cooperation with Russia—Foreign Minister. *UrduPoint.* Accessible at: https://www.urdupoint.com/en/world/rpt-south-africa-ready-to-discuss-nuclear-e-549632.html.

Rogerson, C. M. (2010). Local economic development in South Africa: Strategic challenges. *Development Southern Africa, 27*(4), 481–495.

Roopnarain, A., & Adeleke, R. (2017). Current status, hurdles and future prospects of biogas digestion technology in Africa. *Renewable and Sustainable Energy Reviews, 67,* 1162–1179.

Rose, P. (2013). Long-term sustainability in the management of acid mine drainage wastewaters—Development of the Rhodes BioSURE Process. *Water SA, 39*(5), 582–592.

Saba, A., Pather, R., & Macupe, B. (2018, March 16). Another child dies in a pit latrine. *Mail & Guardian.* Accessible at: https://mg.co.za/article/2018-03-16-00-another-child-dies-in-a-pit-latrine.

Sachs, J., & Warner, A. (1995). *Natural resource abundance and economic growth* (Development Discussion Paper No. 517a). Cambridge, MA: Harvard Institute for International Development.

Sachs, J., & Warner, A. (2001). The curse of natural resources. *European Economic Review, 45*(4–6), 827–838.

SADHS. (2017, May). *South Africa Demographic and Health Survey 2016 key indicators report 03-00-09.* Pretoria: Statistics South Africa. Accessible at: https://dhsprogram.com/pubs/pdf/PR84/PR84.pdf.

SADOE. (2003). *South African Department of Energy, White paper on renewable energy.* Accessible at: https://www.energy.gov.za/files/policies/whitepaper_renewables_2003.pdf.

SADOE. (2011). *South African Department of Energy, Integrated Resource Plan for electricity, 2010–2030.*

SADOE. (2018). *South African Department of Energy, Draft Integrated Resource Plan for electricity, 2018–2040.*

SALGA. (n.d.). *Local government energy efficiency and renewable energy strategy: Full strategy document.* Cape Town. Accessible at: http://sustainable.org.za/sustainable.org.za/uploads/files/file100.pdf.

SAMI. (2017, October 24). *South Africa's coal exports: Where is it going?* Blog. Accessible at: https://www.southafricanmi.com/blog-24oct2017.html.

SAMI. (2018). *GDP growth calculator.* South African Market Insights. Accessible at: https://www.southafricanmi.com/gdp-growth-estimator.html.

Sasol. (2017). *Overview.* Accessible at: https://web.archive.org/web/20170831214232/http://www.sasol.com/about-sasol/strategic-business-units/energy-business/overview.

Satgar, V. (Ed.). (2018). *The Climate Crisis: South African and Global Democratic Eco-Socialist Alternatives.* Johannesburg: Wits University Press.

Scarse, I., & Smith, A. (2009). The non-politics of managing low carbon socio-technical transitions. *Environmental Politics, 18*(5), 707–726.

Scholvin, S. (2014). South Africa's energy policy: Constrained by nature and path dependency. *Journal of Southern African Studies, 40*(1), 185–202. https://doi.org/10.1080/03057070.2014.889361.

Shaw, J. (2017, March). *Assessing the sustainability of an independent power producer's social investment in a community: A case study of Scatec Solar* (MPA dissertation). Stellenbosch University, Stellenbosch.

Shove, E., & Walker, G. (2007). CAUTION! Transitions ahead: Politics, practice, and sustainable transition management. *Environment and Planning A, 39*(4), 763–770.

Smil, V. (2006). *Energy.* Oxford: OneWorld.

SARVA. (2018). *South African Risks and Vulnerability Atlas.* Accessible at: http://sarva2.dirisa.org/.

SourceWatch. (2018a). *Medupi Power Station.* Accessible at: https://www.sourcewatch.org/index.php/Medupi_Power_Station.

SourceWatch. (2018b). *Kusile Power Station.* Accessible at: https://www.sourcewatch.org/index.php/Kusile_Power_Station.

South Gauteng High Court. (2016). *Nkala and Others v Harmony Gold Mining Company Limited and Others*. Accessible at: http://www.saflii.org/za/cases/ZAGPJHC/2016/97.html.

Sovacool, B. (2008, August). Valuing the greenhouse gas emissions from nuclear Power: A critical survey. *Energy Policy, 36*(8), 2940–2953.

Sparks, D., Madhlopa, A., Keen, S., Moorlach, M., Dane, A., Krog, P., et al. (2014). Renewable energy choices and their water requirements in South Africa. *Journal of Energy in Southern Africa, 25*(4), 80–92.

Smith, A., Stirling, A., & Berkhout, F. (2005). The governance of sustainable socio-technical transitions. *Research Policy, 34*(10), 1491–1510.

Smith, C. (2018, March 1). Drought impact on W. Cape economy worse than anticipated—Minister. *Fin24*. Accessible at: https://www.fin24.com/Economy/drought-impact-on-w-cape-economy-worse-than-anticipated-minister-20180301.

Solomons, I. (2017, February 6). Domestic coal sales becoming more important than export market—Prevost. *Mining Weekly*. Accessible at: http://www.miningweekly.com/article/domestic-coal-sales-becoming-more-important-than-export-market-prevost-2017-02-06.

Starr, C. (1999). Observations on the future of nuclear power and how to get there. In B. Kursunoglu, S. Mintz, & A. Perlmutter (Eds.), *Preparing the ground for renewal of nuclear power* (pp. 29–34). Boston, MA: Springer.

StatsSA. (2017). *Non-financial census of municipalities report, 2017*. Accessible at: http://www.statssa.gov.za/?p=11199.

Steenkamp, L., Lategan, R., & Raubenheimer, J. (2016). Moderate malnutrition in children aged five years and younger in South Africa: Are wasting or stunting being treated? *South African Journal of Clinical Nutrition, 29*(1), 27–31.

Stefan, A., & Paul, L. (2008). Does it pay to be green? A systematic overview. *Academy of Management Perspectives, 22*(4), 45–62.

Stockhammer, E. (2015). Rising inequality as a cause of the present crisis. *Cambridge Journal of Economics, 39*(3), 935–958.

Stoddard, E. (2017, November 7). Deaths spike in SA's deep and dangerous mines, reversing trend: Ends nine straight years of declining fatalities. *Reuters* via *Moneyweb*. Accessible at: https://www.moneyweb.co.za/news-fast-news/deaths-spike-in-sas-deep-and-dangerous-mines-reversing-trend/.

Styan, J. (2015). *Blackout: The Eskom crisis*. Cape Town: Jonathan Ball.

Swilling, M., & Annecke, E. (Eds.). (2012). *Just transitions: Explorations of sustainability in an unfair world*. Cape Town: UCT Press.

Swilling, M., Musango, J., & Wakeford, J. (2015). Developmental states and sustainability transitions: Prospects of a just transition in South Africa. *Journal of Environmental Policy & Planning*. https://doi.org/10.1080/1523908x.2015.1107716.

Tait, L., Wlokas, H. L., & Garside, B. (2013). *Making communities count: Maximising local benefit potential in South Africa's Renewable Energy Independent Power Producer Procurement Programme (RE IPPPP)*. Cape Town: International Institute for Environment and Development.

Tcherneva, P. R., & Wray, L. R. (2005, December). *Gender and the job guarantee: The impact of Argentina's Jefes program on female heads of poor households* (Working Paper No. 50). Kansas City, MO: Center for Full Employment and Price Stability. Accessible at: http://www.cfeps.org/pubs/wp-pdf/WP50-Tcherneva-Wray.pdf.

Theron, A. (2018, February 5). Copper theft remains a serious concern—Eskom. *ESI Africa*. Accessible at: https://www.esi-africa.com/copper-theft-remains-a-serious-concern-eskom/.

Thomas, S. (2009, June 22). The demise of the Pebble Bed Modular Reactor. *Bulletin of the Atomic Scientists*.

Trading Economics. (2018). *South Africa: GDP growth rate*. Accessible at: https://tradingeconomics.com/south-africa/gdp-growth-annual.

Tshwane, T. (2018, October 4). Ramaphosa's jobs summit plan to create 275, 000 new jobs annually. *Mail & Guardian*. Accessible at: https://mg.co.za/article/2018-10-04-ramaphosas-jobs-summit-plan-to-create-275-000-new-jobs-annually.

Turton, A. (2009). South African water and mining policy: A study of strategies for transition management. In D. Huitema & S. Meijerink (Eds.), *Water policy entrepreneurs: A research companion to water transitions around the globe* (pp. 195–214). Cheltenham, UK: Edward Elgar.

UBS Investment Research. (2013, January 2). *Entergy Corp.: Re-assessing cash flows from the Nukes*.

Umeozor, E. C., Jordaan, S. M., & Gates, I. D. (2018). On methane emissions from shale gas development. *Energy, 152*, 594–600.

UN. (1957, July 8). *Treaty No. 4234 between the United States of America and the Union of South Africa*. Agreement for co-operation concerning the civil uses of atomic energy. Washington, DC. Accessible at: https://treaties.un.org/doc/Publication/UNTS/Volume%20290/volume-290-I-4234-English.pdf.

UNDESA. (2014). *International decade for action 'Water for Life' 2005–2015*. Accessible at: http://www.un.org/waterforlifedecade/scarcity.shtml.

Valodia, I., & Francis, D. (2016, November 28). How the search for a national minimum wage laid bare South Africa's faultlines. *The Conversation*. Accessible at: https://theconversation.com/how-the-search-for-a-national-minimum-wage-laid-bare-south-africas-faultlines-69382.

Van den Berg, J. (2013). Submission to NERSA: Eskom MYPD 3 Application. *South African Renewable Energy Council*. Accessible at: http://www.nersa.org.za/Admin/Document/Editor/file/Consultations/Electricity/Presentations/South%20African%20Renewable%20Energy%20Council.pdf.

Van Vuuren, H. (2006, May). *Apartheid grand corruption: Assessing the scale of crimes of profit from 1976 to 1994*. A report prepared by civil society in terms of a resolution of the Second National Anti-Corruption Summit for presentation at the National Anti-Corruption Forum. Cape Town: Corruption and Governance Institute for Security Studies.

Van Wyk, J. J., Rademeyer, J. I., & Van Rooyen, J. A. (2010, November). *Position statement on the Vaal River system and acid mine drainage*. Department of Water Affairs. Accessible at: http://www.dwaf.gov.za/Projects/AMDFSLTS/ Documents/Vaal%20River%20System%20&%20AMD%20Version%203.pdf.

Vardi, I. (2016, March 7). Hillary Clinton Showed Support, Associates Profited from Ex-Im Bank Financing World's Largest Coal Plants in South Africa. *Desmog*. Accessible at: https://www.desmogblog.com/2016/03/07/hillary-clinton-showed-support-associates-profited-building-world-s-largest-coal-plantssouth-africa.

Voosen, P. (2018, April 26). Second-largest earthquake in modern South Korean history tied to geothermal plant. *Science*. Accessible at: http://www.sciencemag.org/news/2018/04/second-largest-earthquake-modern-south-korean-history-tied-geothermal-plant.

Wakeford, J. (2012). *Socioeconomic implications of global oil depletion for South Africa: Vulnerabilities, impacts and transition to sustainability* (PhD thesis). Stellenbosch University, Stellenbosch.

Walwyn, D. R., & Brent, A. C. (2015). Renewable energy gathers steam in South Africa. *Renewable and Sustainable Energy Reviews, 41*, 390–401.

Ward, C. (2018, March 14). Biofuel from sugarcane—Why is SA not rushing ahead? *Daily Maverick*. Accessible at: https://www.dailymaverick.co.za/article/2018-03-14-op-ed-biofuel-from-sugarcane-why-is-sa-not-rushing-ahead/.

Wassung, N. (2010). *Water scarcity and electricity generation in South Africa* (MPhil thesis). School of Public Management and Planning, University of Stellenbosch. Accessible at: http://hdl.handle.net/10019.1/18158.

WaterAid. (2017). Out of Order The State of the World's Toilets 2017. Accessible at: http://www.wateraid.se/dokument/Out_of_Order_report_2017.pdf.

Webster, E. (2017, December 11). South Africa needs a fresh approach to its stubbornly high levels of inequality. *The Conversation*. Accessible at: https://theconversation.com/south-africa-needs-a-fresh-approach-to-its-stubbornly-high-levels-of-inequality-87215.

Weisskopf, T. E. (1979, December). Marxian CRISIS theory and the rate of profit in the postwar U.S. economy. *Cambridge Journal of Economics, 3*(4), 341–378.

Western Cape Province. (2019). *Upgrade to a grid-tied solar PV system*. Accessible at: https://www.westerncape.gov.za/energy-security-game-changer/upgrade-grid-tied-solar-pv-system.

Wheels24. (2017, May 4). SA fuel price breakdown: Here's all you need to know. *Wheels24*. Accessible at: https://www.wheels24.co.za/AutoTrader/sa-petrol-price-heres-how-much-fuel-will-cost-you-20170405.

WHO. (2017). *World Health Organization, Cholera data*. Accessible at: http://apps.who.int/gho/data/node.main.176?lang=en.

WikiLeaks. (2010). *Eskom and the World Bank loan for Medupi*. Accessible at: https://wikileaks.org/plusd/cables/10PRETORIA125_a.html.

Willis, M., Jusko, T., Halterman, J., & Hill, E. (2018, October). Unconventional natural gas development and pediatric asthma hospitalizations in Pennsylvania. *Environmental Research, 166*, 402–408. https://doi.org/10.1016/j.envres.2018.06.022.

Wilson, E., Stephens, J., & Peterson, T. (2015). *Smart grid (r)evolution: Electric power struggles*. New York: Cambridge University Press.

World Bank. (2017, March 23). *Tackling tuberculosis in Southern Africa's mineworkers with an innovative approach*. Washington, DC: World Bank Group. Accessible at: http://www.worldbank.org/en/news/feature/2017/03/23/tackling-tuberculosis-in-southern-africas-mineworkeres-with-an-innovative-approach.

World Bank. (2018a). *World Bank Group data: Electricity production from renewable sources, excluding hydroelectric (kWh)*. Accessible at: https://data.worldbank.org/indicator/EG.ELC.RNWX.KH?end=2015&start=1992.

World Bank. (2018b). *World Bank Group data: Electricity production from hydroelectric sources (%)*. Accessible at: https://data.worldbank.org/indicator/EG.ELC.HYRO.ZS?locations=ZF.

World Bank. (2018c). *World Bank Group data: Coal*. Accessible at: https://data.worldbank.org/indicator/eg.elc.coal.zs?end=2015&start=1992.

World Bank. (2019). *Access to electricity—South Africa*. Washington, DC: World Bank Group. Accessible at: https://data.worldbank.org/indicator/EG.ELC.ACCS.ZS?locations=ZA.

Wu, G. et al. (2017). Strategic siting and regional grid interconnections key to low-carbon futures in African countries. *Proceedings of the National Academy of Sciences, 201611845*, 1–9. Accessible at: https://doi.org/10.1073/pnas.1611845114.

Yelland, C. (2009). *Independent Power Producers (IPPs) Organise Collectively to Take on Eskom*. EE Publishers.

Yelland, C. (2011, April 28). Eskom, BHP Billiton and the secret electricity pricing deals. *Daily Maverick*. Accessible at: https://www.dailymaverick.co.za/article/2011-04-28-eskom-bhp-billiton-and-the-secret-electricity-pricing-deals/.

Yelland, C. (2015, January 24). Eskom's Boiler contractor at Medupi and Kusile Defends its reputation. *Moneyweb*. Accessible at: https://www.moneyweb.co.za/uncategorized/eskoms-boiler-contractor-at-medupi-and-kusile-defe-2/.

Young, O. (1994). *International governance: Protecting the environment in a stateless society*. Ithaca: Cornell University Press.

INDEX

© The Editor(s) (if applicable) and The Author(s) 2020   175
A. Lawrence, *South Africa's Energy Transition*, Progressive Energy
Policy, https://doi.org/10.1007/978-3-030-18903-7